Structural adjustment

The International Monetary Fund and the World Bank are key players in the global economy, but their role often goes unnoticed. The Structural Adjustment Programmes (SAPs) they introduce in the wake of financial crises aim to reform economies and ensure a flow of revenue for debt repayment. While this economic management seems sensible, it can have extremely harsh effects which can undermine development.

Structural Adjustment: Theory, Practice and Impacts examines the genesis of these problems and reveals the damaging impacts that SAPs can have. Starting with a look at how the debt crises of the 1970s forced developing countries to seek external help, the book moves on to review what constitutes a standard adjustment programme, detailing the political, economic, social and environmental impacts of SAPs. The final section draws together theoretical critiques of adjustment, as well as political responses on the ground, and presents a case for alternatives to the programmes.

Focusing on Africa, Latin America and Asia, this book presents the origins, impacts and alternatives to SAPs in an accessible and systematic manner and demonstrates the complex effects of these programmes.

Giles Mohan is a Senior Research Fellow in the Department of Geography, University of Portsmouth; **Ed Brown** is a Lecturer in the Department of Geography, Loughborough University; **Bob Milward** is a Senior Lecturer in Economics at the University of Central Lancashire; and **Alfred B. Zack-Williams** is a Reader in Sociology at the University of Central Lancashire.

Structural adjustment
Theory, practice and impacts

**Giles Mohan, Ed Brown,
Bob Milward and
Alfred B. Zack-Williams**

London and New York

First published 2000
by Routledge
11 New Fetter Lane, London EC4P 4EE

Simultaneously published in the USA and Canada
by Routledge
29 West 35th Street, New York, NY 10001

Routledge is an imprint of the Taylor & Francis Group

The right of Giles Mohan, Ed Brown, Bob Milward and Alfred B. Zack-
Williams to be identified as the Author of this Work has been asserted by
them in accordance with the Copyright, Designs and Patents Act 1988.

Typeset in Galliard by Taylor & Francis Books Ltd
Printed and bound by Antony Rowe Ltd, Eastbourne

British Library Cataloguing in Publication Data
A catalogue record for this book is available from the British Library

Library of Congress Cataloging in Publication Data
Structural adjustment theory, practice and impacts / Giles Mohan, ...[et al.]
    p.cm.
    Includes bibliographical references and index.
    1. Structural adjustment (Economic policy)–Developing. I. Mohan,
    Giles, 1966–
    HC59.7.S87368 2000
    339.5'09172'4–dc21   99–048397

ISBN 0–415–12521–9 (hbk)
ISBN 0–415–12522–7 (pbk)

Contents

Illustrations

Figures

Tables

Contributors

Giles Mohan is a Senior Research Fellow in the Department of Geography at the University of Portsmouth. He has researched in West Africa and published most recently in *The Review of African Political Economy*, *Political Geography* and *Space and Polity*. He is currently writing a text on post-colonial African politics entitled *Ripping Yarns* (Addison, Wesley, Longman) and co-authoring a text called *Development, Space and Power* (Sage).

Ed Brown is a Lecturer in the Department of Geography at Loughborough University. His research interests revolve around the changing environment facing 'Southern' economies, especially the ways in which neo-liberalism has asserted itself within contrasting political cultures. Empirical research has been in the Central American region, particularly focussing on the revolutionary process in Nicaragua. Recent publications have appeared in *Political Geography*, *Journal of Historical Geography* and *Geoforum*.

Bob Milward is a Senior Lecturer in Economics at the University of Central Lancashire. He was educated at Ruskin College, Oxford and Jesus College, Cambridge, and is co-author of *Economic Policy* (Macmillan, 1996) and *Applied Economics* (Macmillan, 1998). He is the author of *Marxian Political Economy: Theory, History and Contemporary Relevance* (Macmillan, 2000).

Alfred B. Zack-Williams is a Reader in Sociology at the University of Central Lancashire, and a member of the EWG of the ROAPE. He is the author of several articles and reviews on the political economy of underdevelopment in Africa and the Diaspora, as well as *Tributors, Supporters and Merchant Capital* (Avebury, 1995). He is currently working on child soldiers in the civil wars in Sierra Leone and Liberia.

Preface and prospectus

> One might think, among other things, that in a period of capitalist triumphalism
> there is more scope than ever for the pursuit of Marxism's principal project, the
> critique of capitalism. Yet the critique of capitalism is out of fashion. ... There
> can never have been a time since Marx's day when such a task needed doing
> more urgently, as more and more branches of knowledge, both in cultural studies
> and in the social sciences, are absorbed into the self-validating assumptions of
> capitalism or at least into a defeatist conviction that nothing else is possible.
>
> (Meiksins Wood 1995: 1 and 4)

While teaching Development Studies in Preston a student came to me after a
lecture and asked what single text I could recommend that covered the theories
underlying Structural Adjustment Programmes (SAPs) as well as their various
impacts. I thought for a while and mentioned a couple of chapters and a
pamphlet. As soon as she had left I locked my office and hurtled around to
Alfred B. Zack-Williams' room in the adjacent building. On asking him the
same question that the student had asked me he was equally stumped. It was
from there that this book emerged. After too many years of writing I am
pleased that there is still nothing on the market which covers this range of
material.

But beyond pedagogical and marketing reasons we feel that this book is
important for wider political motivations. As the quote above suggests, the need
to analyse and criticise global capitalism is now more pressing than ever. It is
ironic, but perhaps not coincidental, that at the very time that such a critique was
needed, the one body of theory and praxis which has consistently exposed the
contradictions of capitalism was seen to be in an 'impasse' (Rapley 1994;
Schuurman 1993). This body of theory is, of course, Marxism, which from the
mid-1980s (Booth 1985) onwards began to be discredited within Development
Studies in particular and social theory more generally. The ending of the Cold
War in the late 1980s signalled the practical limitations of actually-existing
socialism and with it the Marxist project.

The triumphalism that greeted the ending of the Marxist alternative and the
hegemony of a globalising capitalism was summed up in Francis Fukuyama's
(1992) *End of History* thesis. He famously argued that liberal democracy allied

to market capitalism was the pinnacle of human achievement (despite being rather dull) and that all contending ideologies were invalid. Subsequent conflicts and crises have attempted to reconcile this assertion by reducing these problems to examples of 'irrationality', in much the same way as modernisation theorists had done in the early days of the Cold War. For example, Sierra Leone was, according to Kaplan (1994), seen to epitomise the anarchic outpourings of uncivilised people. Likewise the currency speculations of Nick Leeson were seen to be the machinations of a single 'rogue' trader who flouted the rules. The overall picture was one of a 'normal' and inevitable logic which dissenting governments ignored at their peril (Agnew and Corbridge 1995).

Underlying this discourse about the 'natural order' of globalisation and free-marketeering was a set of political and economic processes which went largely unseen by the majority of the developed world, but could hardly be ignored by those people living in the Third World. Contrary to Fukuyama and others (for example, Ohmae 1990; Friedman and Friedman 1980), the globalisation of free markets and neo-classical order was anything but natural and inevitable. Huge political and economic resources, known as structural adjustment programmes, were mobilised to ensure that these dissident countries danced to the same tune. It is often overlooked that in the heartlands of neo-liberalism equally heavy-handed and repressive political power was used to weaken the power of labour and remove the state from large sectors of economic and social life. Either way, the neo-liberal world order has been, and is being, created through the intervention of key institutions such as the World Bank and International Monetary Fund. It is ironic that a theory that posits the 'freedom' of markets and the limited use of state power requires massive amounts of political interference to do so (Polanyi 1960).

What is equally ironic, but not at all surprising, is that once serious crises emerged which threatened the growth of the developed world, these issues become worthy of attention in the Western media. For a decade and a half, from the early 1980s, structural adjustment was relatively insidious, taking place in distant and impoverished places that were always in trouble and usually too complicated to comprehend. Once these distant places began to melt down (Russia, Indonesia, Brazil) in the mid-1990s and threaten 'our' development we began to take note of the IMF and its possible role in 'stabilising' and 'adjusting' these economies. It is for these reasons that it is important to understand the concept of 'adjustment' and its contradictory effects.

So this book seeks to expose the evolution of these interventions and question their inevitability. We wanted to show how Third World countries have been 'hemmed in' through colonialism, the Cold War and the debt crisis and how these set the scene for massive interventions in the 1980s and 1990s. The book also reveals how the application of neo-classical orthodoxy created a whole set of new problems or exacerbated existing ones. While artificial, we have separated these into economic, social, political and environmental impacts. Finally, we wanted to show that these interventions have been challenged in both reformist and radical ways and that there are, despite the countervailing discourse, genuine alternatives. Crucially, these alternatives are not just for the hitherto 'recipients' of the neo-liberal medicine in the Third World, but are also questions which 'we' need to ask ourselves.

We and the publishers felt that these issues could only be addressed through an inter-disciplinary approach. As will become clear the scale and scope of adjustment programmes is so great that a single approach is inadequate. In particular the reductionist assumptions of the neo-classical model upon which SAPs rest are so limiting that they overlook many of the unpriced and unseen effects of adjustment. To do justice to this complexity we bring to this book the insights of sociology, economics and geography. While academic boundaries are obviously artificial we feel that each of us contributed something different to the book and we hope that together these insights provide the reader with a comprehensive overview of this important topic. Not only are our disciplinary roots different, but so too are our area specialisms. One of us works on Central America, another on economic theory and policy and the remaining two on West Africa. Initially the Latin Americanist was not part of the project but on the recommendations of an anonymous reviewer we brought him in to counter the Africanist bias since the book is about SAPs in general.

In spite of these differences between us, which we found to be creative and productive, we share a number of common concerns about the ways in which SAPs should be theorised and analysed. These are:

- SAPs need to be set against a critical history of the world economy focusing upon relationships between 'developed' and 'developing' countries;
- SAPs need to be set within a framework of theories and discourses about development rather than being analysed in atomistic, technocratic terms;
- SAPs are spatially specific and differ according to 'local' contexts. They should not simply be treated as monolithic despite the universal assumptions which underpin them. This applies to both the formulation and impacts since states, markets and societies are locally contingent;
- SAPs are also temporally specific and have changed in composition in the light of criticism and practical application;
- SAPs, and the neo-liberal ideas that they reflect, are not the only development pathway. Alternatives do exist and are being actively contested on a daily basis.

First, we strongly believe that SAPs should be set within a longer history of the world capitalist economy. If we were to believe the analysis of the architects of structural adjustment, the economic crises of the 1970s that developing countries suffered can largely be attributed to the failures of *national* policies. They argue that governments adopted the wrong policies following the oil crises of the early 1970s. Such policy failure was exacerbated, according to the same analysts, by a wider problem of 'corruption' and mismanagement. Hence, the target for SAPs became the ways in which national economic policies could promote export-based growth and, most importantly, introduce fiscal probity to allow debt repayment to occur.

What this 'orthodox' analysis fails to analyse is the ways in which developing countries have been consistently tied into the world economy on unequal terms. From the colonial period through to decolonisation and the Cold War and finally the debt crisis, these countries have been marginalised politically and economically so that maldevelopment and crisis are structurally embedded in their political economies. To simply start the clock in the 1970s and blame national mismanagement for the balance-of-payments crises misses out this longer-term struggle over the incorporation of developing countries into the world economy. However, we are not seeking to exonerate the political leaders of many developing countries and heap all the blame on an abstract and venal 'world system'. Some of these leaders did, at times, pursue poor policies which exacerbated other structural problems. Only by appreciating the interplay of these factors in specific contexts can we understand the pressing need to 'adjust' in the late 1970s and early 1980s.

Second, we feel that SAPs should be set within an appreciation of development theories and discourses. Clearly, ideas about the world both shape and are shaped by that same world so it is vital to understand how these ideas change and influence concrete actions. Again, the designers of SAPs take a partial view of the causes and cures of economic crises. Without pre-empting discussion in the book, these decision-makers believe in the power of markets to deliver economic and social justice. They considered, at least in the early days of adjustment, the state to 'distort' markets and so concluded that it should be removed as far as possible from economic life. Such theories saw a return to fundamentalist beliefs about economic equilibrium which have been termed neo-liberal since they advocate the 'freedom' of market-based decisions. In cementing this new orthodoxy the neo-liberals had to discredit contending theories and policies. While a Marxist alternative was reasonably unrealistic, the neo-liberals had to attack other 'statist' models of development, especially those built around some notion of Keynesian market regulation. Having established the unassailable and 'natural' efficacy of the market system, the way was opened for reams of technocratic economic analyses which sought to improve the efficiency of policies designed to 'liberate' markets from the fetters of state regulation and the dampened entrepreneurial spirit of Third World citizens.

Third, we see a tension between the universalising and reductive assumptions of the neo-liberal policy package and the complex realities of the contexts in which it is applied. We have seen that the architects of adjustment blamed misguided national policies for economic crisis whereas a structuralist neo-Marxist interpretation would blame the world system. Neither position is wholly correct nor wholly incorrect. We feel that each national (and local) context is different and that the configuration of state, markets and society is specific to each situation. For example, it would be dangerous to treat the economy of India in the same way as, say, a small, landlocked country such as Burkina Faso.

This realisation has two related implications. In terms of policy formulation we need to endow each national government with some political agency. Arguments which simply state that the lenders forced SAPs on recipient governments underplay the complex ways in which domestic political forces, however limited and unrepresentative, were galvanised around the adjustment agenda. Having said this, we must be aware that recipient governments were severely constrained in their range of options and the lenders were quite heavy-handed in their application of the 'standard' package so that most SAPs look very similar. The other ramification of these contextual differences is the way in which the SAP package impacts upon local conditions. Given the contextual nature of the state–market–society mix, a standard package is likely to have very different effects in practice. This theme recurs throughout the book whether it be in economic, social, political or environmental terms.

Fourth, we believe that SAPs are not only spatially specific, but that the temporal dimension is important. The more structuralist-inspired critiques of SAPs tend to treat them as monolithic in terms of both space and time. While we would not wish to claim that SAPs have radically altered over the past two decades, it is equally ill-advised to argue that they have remained the same. The early orthodoxy was radically anti-state and advocated the market mechanism with a rabid insistence. It must be remembered that this coincided with the unfolding of the harsh Reaganomic and Thatcherite projects which weakened the power of labour and saw the privatisation of many state assets. After the repeated failure of SAPs and the realisation that the state could play a positive role, the neo-liberals softened their anti-statism. Recent moves have seen the incorporation of various 'partnerships' between state and market, between state and society and between market and society. Such 'governance' initiatives, while signifying a changing approach, are still firmly anchored within a belief that markets, albeit imperfect, are the best mechanism for growth and poverty alleviation.

Fifth, and finally, we feel that while the neo-liberal orthodoxy has undergone some piecemeal reflection on its project, there are more far-reaching alternatives to their model. The striking feature of discourses around globalisation, marketisation and liberalisation is that these processes are presented as necessary and inevitable for prosperity. We argue, especially in the last section, that alternatives have been tried in both a formal and informal sense. People have criticised adjustment and called for reforms to existing programmes. In some cases the lenders have incorporated these into subsequent policy design or, more usually, bolted them on as an after-thought in an attempt to remedy the disastrous effects of another policy element. Beyond these more formalised critiques are the lived struggles of people across the globe as the neo-liberal logic becomes more deeply embedded into everyday lives. As the Meiksins Wood quote at the beginning of this preface suggests, more and more areas of our lives are subject to a unifying logic. The insidious nature of this process throws up everyday forms of resistance and struggle which are gradually coalescing into a broader critique of globalising capitalism. It is this unification of life experiences that creates the more intense (and likely) need of a counter-hegemonic project.

These five themes run throughout the pages of this book. Some chapters stress one or more of the themes in greater detail than others. The book is structured into three parts, each of which has an introduction which outlines the individual chapters which follow. The first part examines the history of the world economy to show how developing countries were 'hemmed in' and that, within this logic, accepting the neo-liberal medicine was inevitable. It also details the neo-classical model which underpins the adjustment process. Part II examines the impacts of SAPs and is divided into economic, social, political and environmental considerations. The final part is extensive and reviews the range of 'alternatives' to adjustment.

Giles Mohan

Abbreviations

APEC	Asia–Pacific Economic Co-operation
ASEAN	Association of South East Asian Nations
BWIs	Bretton Woods Institutions
CEM	Country Economic Memorandum
CEPAL	Comision Economica para America Latina (or ECLA)
ECLA	Economic Commission for Latin America (or CEPAL)
EFF	Extended Fund Facilities
ESAF	Enhanced Structural Adjustment Facility
GATT	General Agreement on Tariffs and Trade
GDP	Gross Domestic Product
GEF	Global Environment Facility
GNP	Gross National Product
IBRD	International Bank for Reconstruction and Development
IFC	International Finance Corporation
IFIs	International Financial Institutions
ILO	International Labour Organisation
IMF	International Monetary Fund
ISI	Import Substitution Industrialisation
MAI	Multilateral Agreement on Investments
MOSOP	Movement for the Survival of the Ogoni People
NACLA	North American Congress on Latin America
NAFTA	North American Free Trade Agreement
NEAPs	National Environmental Action Plans
NGOs	Non-Governmental Organisations
NICs	Newly Industrialised Countries
OAU	Organisation of African Unity
OECD	Organisation for Economic Co-operation and Development
OPEC	Organization of Petroleum Exporting Countries
PAIT	Programme of Support and Temporary Income
PAMSCAD	Programme of Action to Mitigate the Social Costs of Adjustment
PNDC	Provisional National Defence Council

RCT	Rational Choice Theory
SAF	Structural Adjustment Facility
SALs	Structural Adjustment Loans
SBAs	Stand-by Arrangements
SAPs	Structural Adjustment Programmes
SDA	Social Dimensions of Adjustment
SDRs	Special Drawing Rights
TNCs	Trans-national Corporations
UN	United Nations
UNCTAD	United Nations Conference on Trade and Development
UNDP	United Nations Development Programme
UNECA	United Nations Economic Commission for Africa
UNICEF	United Nations Children's Fund
WTO	World Trade Organisation

Part I

Histories and theories of adjustment

Introduction

> If development studies means anything it presumably means the study of those countries which have a past in common, if not a common past – I refer to the history of colonialism and post-colonialism – and which can only reasonably be understood in relation to the so-called developed countries.
>
> (Corbridge 1995: x–xi)

In this part we seek to introduce structural adjustment programmes in two ways. The first chapter is historical and demonstrates the ways in which SAPs represented one of the few viable alternatives open to developing countries in the face of economic crises. It also shows that these crises were neither inevitable nor due simply to 'internal' mismanagement, but part of a longer process which tied developed and developing countries into, sometimes exploitative, relationships. We have done this to give a context against which the political and economic 'logic' of SAPs was set as they sought to respond selectively to the causes of these interlocking crises. We say 'selectively responded' since the blame for these crises has usually been placed at the door of the governments of developing countries for adopting 'inappropriate' policies rather than the global system which structures and constrains their economic and political position within it.

The second chapter introduces SAPs by setting out what actually constitutes an 'average' programme as well as explaining the lenders' theoretical premises. We have done this since it underpins part II, in particular, which focuses upon the impacts. Most SAPs are built around a standard package of policy measures designed to address the balance-of-payments position and further integrate national economies into the world system. Given that we analyse their impacts in subsequent chapters, it is vital that the reader knows what each policy aims to do and how it interacts with, and sometimes contradicts, other policy measures. The chapter also shows the economic reasoning behind these policies which is firmly rooted in neo-classical thought. The chapter concludes with a critique of these theoretical models, at the level of theory, while the following part criticises their articulation in practice.

1 The long road to structural adjustment

Alfred B. Zack-Williams with Ed Brown and Giles Mohan

Introduction

In this chapter we argue that structural adjustment programmes (SAPs) must be seen in the context of the post-independence development experience. In particular we stress that various processes of underdevelopment set the conditions in which these programmes could be so radically implemented. In doing this we emphasise that countries undergoing SAPs have different experiences of capitalist development so that we can generalise neither about the 'causes' nor, as part II argues, the impacts of adjustment. Indeed, the final part on alternatives argues that such responses to adjustment must also be contextually specific. Having said this, one of the major features of all SAPs is their underlying logic based, as we see in chapter 2, on neo-classical principles which were revitalised in the late 1970s and have become the ascendant ideology of the late twentieth century.

The chapter is historical and charts the decline of Third World countries from their optimistic hopes at independence to the crisis that precipitated adjustment. Both 'internal' and 'external' factors will be examined, although this distinction is somewhat artificial since the process is one of mediation and mutual definition. However, it is important to continually emphasise, either through case studies or regional analyses, that while certain experiences were common there are significant differences between countries and regions and across time spans. For example, whilst adjustments in most African countries led to growing political authoritarianism and a weak labour movement (Zack-Williams 1997), in the Dominican Republic and Jamaica SAPs led to 'democratic renewal' as implementing regimes were removed from office through the democratic process (Espinal 1995; Bernal 1984).

Robert Cassen reminds us that 'before structural adjustment, there was international adjustment' (Cassen 1994: 7–14). Similarly, pointing to the continuity in the use of the term, Toye notes that prior to the 1980s, it had been used by the developed world to transform their old industries, such as textiles, which could no longer compete with those of the newly industrialising countries. However, by the early 1980s, there had been a major shift of the 'locus of responsibility for ensuring successful world development, from the shoulders of the economically strong to those of the economically weak' (Toye 1994: 19). This shift, and the sub-sequent changes in the global economy, as well as within Third World societies after the 1970s, saw the gradual process of economies and societies being 'fixed'. This dialectical process of, on the one hand, being 'hemmed in' (Callaghy and Ravenhill 1993) and, on the other, being marginalised from the central production processes of global capitalism is the direct result of adjustment programmes.[1] By the middle of the 1980s, SAPs had taken over as the main tool for correcting macro-economic imbalances, which were seen as obstacles to sustained growth. For example, 'By the end of the decade, 32 of the 45 countries in sub-Saharan Africa had been involved in a structural adjustment programme' (Roth 1991: 31); and in Central America between 1980 and 1988 twelve emergency agreements were signed with the IMF (Dolinsky 1990). In this chapter, we want to locate some of the problems, which impelled Third World economies on to the long route to structural adjustment programmes.

The post-colonial state and the imperative for reform

A major feature of the post-colonial state is that, like its precursor, the colonial state, it is the creature of imperialist forces, designed to serve the interests of non-domestic classes. However, colonisation took place over many years and in different contexts so it is inadvisable to treat all countries and regions as homogenous. In this section we review briefly the experiences of different regional economies since independence and see how the events of the 1970s, in particular the oil crises, pushed most countries into deep recession and initiated the debt crisis and the subsequent need for multilateral borrowing.

Africa

By the end of the colonial period in Africa, merchant capital had emerged as the dominant form of capital (Kay 1975), and this was typified by the preponderance of metropolitan trading agencies. Not only did the colonial trading houses dominate the produce trade, but they also owned the ships on which these goods were transported, and some were involved in wholesale retail trades. At the outbreak of war in 1939, the colonial governments granted monopsony power to the produce marketing boards to purchase and export produce from peasant producers in each territory, and the proceeds were held in British government security in London. The various independent governments continued with this policy until the advent of structural adjustment programmes in the 1980s. Those Africans in the retail trade sector were now being squeezed by new competitors, the Levantine traders (Hopkins 1973). The failure of the colonial financial institutions to offer loans to Africans tended to exacerbate their vulnerable position (Zack-Williams 1995a).

With the exception of Latin America, most Third World states gained independence during the Cold War era when super-power rivalries were being played out in various theatres, especially in the Third World. As these nations

tried to play off one super-power against another, they sought to extract concessions and preferential economic relations. On the political front there was much
optimism about Third World development, as reflected in the growth of 'Third
Worldism' and political movements such as the Non-aligned Movement, pan-
African movements, and the Afro-Asian Solidarity Movement.

As we have seen the leaders of these nations followed various economic and
ideological paths to development. In Africa at least, the governing classes which
took over the reins of government from the departing colonialists did very little
to shift the economy and polity away from that which was designed by the
rulers of yesteryear. The colonial mode of accumulation was maintained, this
time with the governing classes playing a major clientelistic role. They, unlike
the ruling class of the colonial epoch, are neither innovative nor do they constitute a capital-owning class. They depend on rent extraction for their reproduction
such that in moments of crisis, when rent-seeking activities are reduced, this
poses major threat to the legitimacy of this class for formal control of the state.
Soon the post-colonial state became the pivot of much of the neo-imperial
brokerage so that states became unaccountable and bloated.

Nonetheless, because of its precarious hold on society, the governing classes
failed to mobilise their previous coalition allies, the workers and peasants. In
many African countries, the struggle for political independence was based on a
coalition or 'united front' to confront the colonial power. However, no sooner
was independence gained than the petty bourgeoisie shunned their former
coalition partners, by alienating them from the centre of political and economic
activities. It was not long before the monolith, the one-party state, replaced the
expected dictatorship of the proletariat. In many cases the one-party state led to
'crony-statism',[2] reflecting the 'malleability' and 'institutional fragility' of the
state (Kandeh 1992). Petty-bourgeois rule could not be 'secured [without] the
uncoerced compliance of subordinate strata to their rule' (ibid.: 30).

In much of the Third World and sub-Saharan Africa in particular, the period
1960–1980 witnessed the rise of 'economic nationalism' (Ferguson 1988), that
is the growth in 'policies of substantial regulation and control by planners'
(Grosh 1994: 29) regardless of the different economic paths which these
nations pursued. This was in line with the dominant thinking of Keynesian
welfarism in the West, and of the command economies of Eastern Europe. As
Synge (1989) has observed, countries such as Kenya and Ivory Coast privileged
the fortunes of multinational corporations and export agriculture. Others such
as Ghana and Zambia seized the revenues of exporters and channelled them
into a multitude of government-owned import-substituting firms, whilst
Nigeria's faith in economic nationalism (Synge 1989: 15) meant that '[govern-
ment] control over capital markets [was used] to subsidise investments by
private businessmen and local corporations, thereby promoting the formation of
an indigenous bourgeoisie' (Bates 1994: 15). Bates notes that, despite the
differing 'developmental trajectories', African states were all developmentally
activistic, in order to overturn the market solution.

As noted above, state intervention was not a peculiarity of African policy-makers.

It was quite common with most Third World decision-makers; the difference lies not in the depth of such interventions, but the quality.[3] In many African countries, such as the Afro-Marxist states of Mozambique, Benin and Guinea Conakry, and the African socialist states, there were widespread nationalisations.[4] Even in countries, such as Liberia, Ivory Coast and Sierra Leone with 'open door' economic policies, hostility to foreign investment was not unknown.[5] Seckler and Cobb (1991) contrast Asian leaders with their African counterparts, pointing to African leaders scrambling daily to find budget resources to pay civil servants by contrast with Asian leaders who had followed pragmatic long-term plans varying from five to twenty-five years. Seckler and Cobb also observed that Asian leaders emphasised the importance of economic growth. There was also much emphasis placed on a well-trained and educated labour force by Asian leaders, whilst emphasising macro-economic stability and matching public expenditure in line with public revenue.

Seckler and Cobb drew attention to the fact that whilst many African governments were hostile to the private sector, Asian governments saw this sector as central to growth:

> Some (Asian) countries emphasised import substitution (and protection); some (particularly Korea) used indicative planning and substantial moral suasion to get the private sector to behave the way the state wanted; para-statals were frequently part of the landscape; many had an activist industrial policy; some countries used taxes and subsidies to shift the incentive structure. But the bottom line was that all governments were highly pragmatic, and market prices were used to guide public intervention.
>
> (ibid.: 16)

There were also huge investments in agriculture and rural infrastructure, thereby raising agricultural productivity; and it was 'at a key point in their development that Asian countries turned to exports as the engine of growth' (ibid.: 17). Seckler and Cobb also argued that in Asia, governments were the embodiment of the state, which tended to strengthen their claim to legitimacy. By contrast in much of sub-Saharan Africa, lack of investment in the rural sector resulted in a fall in agricultural productivity, and general poor economic performance leading to the rise of 'parallel markets' and 'parallel states', thus posing a threat to the authority of the governing class.

The resulting abysmal economic performance is reflected in the fall in the GNP for sub-Saharan African states from US$200,080 million in 1980 to US$156,313 million in 1990. By contrast the GNP (in millions) for all other Third World regions such as East Asia and the Pacific rose from US$528,189 to US$874,990; Latin America from US$690,342 to US$1,057,666; South Asia from US$220,757 to US$374,433 for the same period.[6] Thus the 1980s was a period of rapid economic decline for sub-Saharan Africa, whilst other regions in the Third World experienced growth.

In pointing to the positive economic performances of areas like Asia and

Latin America, it is important to note that these areas were not free from economic and political problems as we see below.

It must be pointed out that these aggregate figures mask widespread disparities of performance. For example, the GNP of Botswana rose steadily during the 1980s from US$824.0 million in 1980 to US$2,212.5 million in 1989; however, countries such as Benin, Chad, Mauritania, Kenya, Liberia and Malawi experienced a drop in GNP in the middle of the decade. Others such as Nigeria, Mozambique and Sierra Leone experienced a worsening performance over the decade (see note 6).

Asian economies, in particular, experienced high rates of growth of export earnings in the 1980s, despite deteriorating terms of trade as high as 30 per cent for Indonesia and 26 per cent for Malaysia (Seckler and Cobb 1991). By contrast, most African economies experienced not only deterioration in the unit value of exports, but there was decline in the total volume of some leading exports. For example, the price of cocoa in the international markets dropped from £3,000 in 1977 to £600 in 1986. However, in the case of Sierra Leone, not only was there a deterioration in the unit value of exports, but there was a decline throughout the 1980s in the total volume of some leading export items. Total volume of exported cocoa fell from 12,500 metric tonnes in 1983 to 8,600 in 1986. Similarly, recorded diamond exports dropped from 2 million carats in 1970 to 595,000 carats in 1980 and a derisory 48,000 carats in 1988 (*The Courier* 1989: 23–49). Many African countries lost their pre-eminence as exporters of certain commodities. For example, Ghana ceased to be a leader in the world cocoa market and Nigeria moved from a palm oil exporting nation to an importing one.

This loss in value and volume of exports impacted upon the balance of payments. The situation was not helped by three other economic policies. First, the import-substitution strategy of industrialisation proved to be a failure, since the foreign exchange savings that it was supposed to effect never materialised (Dolinsky 1990; Zack-Williams 1995a). This was due to the fact that there was little or no forward or backward linkage with other domestic sectors, as well as the high import content of this sector. In response, many Third World governments set up infrastructure and other fiscal incentives such as factories and tax holidays in order to attract foreign investors.

Second, as we have noted above, post-colonial African governments continued with the policy of forced savings, by taxing peasant producers through the produce marketing board system, which alienated African rural producers. Third, the policy of running huge budget deficits and an over-valued currency led to rampant inflation in most African economies. In many cases, African governments intervened in the economy not for efficiency reasons, but to provide opportunities for clients and supporters, leading to a bloated bureaucracy.

These three policy actions plus the debilitating effects of the oil price rises ushered in a world recession which lowered demand for primary products of the Third World while industrialised countries raised their protective barriers and oil-importing countries faced major balance-of-payments problems. The huge profits reaped by oil-producing countries resulted in a glut of cheap credit as petrodollars filled the banks, which soon became the source of Third World indebtedness.

Latin America

Independence came earlier to Latin America, when the technology of imperialism of European states was less well developed and the world economy was less integrated. Lehmann (1990) traces Latin American political thought back to a nineteenth-century positivist-modernist nexus which produced a two-fold belief in cultural distinctiveness on one hand (especially with respect to the USA) and a belief in state centralism on the other. It is this genesis that Lehmann argues helps explain the tendency of Latin American countries to adopt import substitution industrialisation (ISI) policies which were state-led and attempted to break from the reliance on imported manufactured goods.

While political culture is certainly a factor in the move towards ISI in Latin America we need also to look at the region's position within wider geo-political and economic relations. Green (1995) argues that Latin America has, since independence in the early nineteenth century, been afflicted by fifty-year economic cycles which led to major economic restructuring, but not the 'creative destruction' envisaged by Schumpeter (1934). In the 1820s the strategy was for production of raw materials for export which Frank (1978) analysed in such a persuasive and politically-charged way (Lehmann 1990). After the subsequent crisis of the 1870s some republics attempted low-key manufacturing, but the major impetus towards ISI was the depression of the 1930s when commodity prices slumped as world demand dropped off.

By then the USA was the ascendent economic power and, as the CEPAL economists noted, the country compared to Britain, the previous hegemon, had a lower propensity to import raw materials. Hence, economic development could not be guaranteed through exporting primary commodities so that some form of industrialisation was necessary. It was then that the ISI policies were put in place. They attempted to promote 'inner-directed' development involving 'the replacement of goods produced domestically; the policy instruments used to achieve this include, principally, government manipulation of the exchange rate, import tariffs or quotas, subsidised credit for substitutive investments and direct or indirect subsidies to hold down the costs of inputs required for substitutive production' (Lehmann 1990: 7). Crucially, this strategy relied heavily on intervention by the state to control trade and investment as well as involving corporatist labour relations with the trade unions being called upon to support these national projects.

The effects of ISI were mixed. The firms were heavily protected from foreign competition and so could effectively 'shelter' without any threat. This led to shoddy manufacturing which not only burdened the consumer with sub-standard goods, but also undermined the longer-term international competitiveness of these firms. The import-substituting firms also operated as *de facto* oligopolies and fixed prices at inflated levels, because there was no effective competition. In addition, although the ISI policies ended the need to import manufactured goods they actually deepened the dependence on developed world economies since they relied upon imported capital goods such as machinery. Over-valued exchange rates exacerbated the problem since these firms could not find external markets for their over-priced, poor-quality goods yet were facing mounting import bills due to the reliance on foreign capital goods.

In class terms the ISI policies widened income inequalities especially between urban and rural areas. Farmers faced price controls while the over-valued currency encouraged the importation of cheap foodstuffs which undercut local farmers. One result was mass out-migration from rural areas and the related phenomenon of hyper-urbanisation. Political leaders were loath to sanction the foreign TNCs operating in their countries while they also believed them to be a source of 'modern' technology. In the late 1960s, the Brazilian and Mexican governments in particular realised that ISI was unravelling and that they needed to seriously promote exports. Brazil, for example, devalued its currency and ruthlessly suppressed labour to ensure low-wage competitiveness. The result was an 'economic miracle'. But given their knowledge, expertise and reach it was the TNCs which were best placed to promote this export drive so that domestic firms benefited little. It was around this time that the structuralist and neo-Marxist dependency critiques emerged as it was becoming increasingly obvious that ISI policies were not weakening ties of dependence upon foreign capital.

Throughout this period debt had been an ever-present feature since domestic markets were often not large enough to sustain many firms. Hence, the funding of capital goods was through loans and a growing trade deficit. As Dolinsky has noted:

> the process of indebtedness in the region has been historically associated with the needs and contradictions which marked the process of modernisation of dependent capitalism. This has been true of the economic process in the region, even during the boom of the 1960s and 1970s.
>
> (Dolinsky 1990: 76)

So it was not surprising that when the oil shocks hit, the economies of Latin America were ill-equipped to cope and were the first countries to declare themselves bankrupt.

East Asia

The tiger economies of East Asia present a somewhat different case. In the post-war period, in particular the 1960s, they grew consistently, even through the oil shocks of the 1970s (Balassa 1981). We noted above some of the salient differences between these successes and the decline of African economies over the same period. However, it is worth examining briefly their political economies as they have also undergone adjustment and major restructurings.

In terms of the success of the NICs the following factors are important

(Bello and Rosenfeld 1990). First, was their 'special relationship' with the United States which saw them tying their hopes to the export of labour-intensive, standardised products to the vigorously expanding US market. During this period the USA championed free trade on the one hand and anti-communism on the other. The place of the NICs as bulwarks against communist threats resulted in huge inflows of US aid and investment. It also meant that while the Bretton Woods Institutions were promoting the worldwide removal of protectionist barriers in the late 1960s, the USA turned a blind eye to the heavy restrictions and subsidies issued by the NIC governments.

Second, was the NICs' close relations with Japan. Japan had been the regional imperial power in the middle part of the twentieth century and left behind a reasonable infrastructure and, more importantly, the legacy of post-colonial ties. When the Japanese economy faced rising labour costs, firms relocated to their former colonies where labour was cheaper, while Japanese trading houses handled regional trading relations. Additionally, a measure of technology transfer occurred from Japan to the NICs so that a virtuous circuit existed between the manufacturing NICs, the technologically advanced Japanese and the burgeoning US markets. The result was that by the mid-1980s the NICs were running large trade surpluses with the USA and huge deficits with Japan which deepened their economic dependence upon their former colonial master (Bello and Rosenfeld 1990).

The third factor concerns the relatively high levels of state intervention, except perhaps in Hong Kong. The political leadership was relatively insulated from civil society and in control of a disciplined and well-educated technocracy (Leftwich 1995). Together they were able to direct economic policy via subsidies, preferential credit arrangements and various investment incentives – all of which were geared around export-promotion. However, the financing of this development path required heavy external borrowing which was to hurt these countries when interest rates rose, but these shocks were off-set by other adjustments.

The oil shocks in the early 1970s hit the NICs hard given that they were highly trade dependent and the industrialised countries raised protections in the aftermath (Chowdhury and Islam 1993). The NICs were also dependent upon oil imports which became increasingly expensive over the decade. However, unlike the African and Latin American economies mentioned above, the NICs suffered less harshly due to concerted adjustment and industrial restructuring (Haggard 1990). Briefly, the NICs undertook 'absorption reducing' policies which involved huge cuts in public expenditure which were possible due to a combination of the insularity of the state, the weakly developed political opposition and outright repression. However, to ensure that this policy did not reduce investment the governments maintained selective credit and finanical policies. The other adjustment mechanism was 'switching' which was achieved via devaluation of the currency. While this should have resulted in labour protests as import costs rose and real wages declined, the strong state and weak unions ensured that any protest was short lived.

In addition to these fiscal and monetary measures there were a set of industrial policies. The first sought to upgrade traditional export sectors such as textiles, steel and electrical components (Cheng and Haggard 1987). Second, they diversified into new product lines which saw a second round of 'import-substitution' in the intermediate and capital goods sectors as well as looking for new opportunities on world markets. Third, the NICs encouraged producer services, especially in Hong Kong and Singapore which are now major financial centres. In Korea and Taiwan this did not occur and recent financial scandals suggest that the banks remained bound closely to political decision-making. Fourth, the NICs sought new geographical markets, especially in attempting to break their dependence upon Japan and the USA. For example, the Middle East has been a major recipient of Korean investment in the construction industries.

After riding out the shocks in the 1970s, the NIC economies saw the unravelling of some of the factors which had given them so much success in the 1960s and 1970s. By the late 1980s the USA was increasingly protectionist, which reduced the main market for exports, while the Americans also forced the appreciation of the Taiwanese and Korean currencies against the dollar in order to make their exports more expensive for US consumers. Internal problems were also growing in the shape of rising labour costs and increasing labour militancy, which saw some relocation of industry to the 'second wave' of NICs as well as the capitalisation of the production process. Additionally, the export push of the previous two decades had seen a neglect of agriculture which resulted in rising food imports, especially from the USA. Allied to this was the erosion of environmental resources in the rush to industrialise (Bello and Rosenfeld 1990).

The debt crisis

By the late 1980s, indebtedness had become a major problem with one country after another defaulting on their payments so that arrears mounted. The situation was not helped by the growing indebtedness to the USA, which led to high interest rates on the loans to Third World countries. In the case of Central America, between 1981 and 1984 the total debt of the region grew at an annual rate of 15 per cent and by 6.3 per cent between 1984 and 1987 (Dolinsky 1990), and by 1987 the foreign debt of the region was in excess of US$19 billion or 89 per cent of the region's GNP (ibid.). Long-term debt servicing consumed over 22 per cent of the region's total export earnings (ibid.).

Table 1.1 offers a picture of the debt situation in various regions of the Third World and it shows that by Third World standards, sub-Saharan Africa's debts were relatively small. None the less, the smallness of the African debt is a reflection of the marginal position of Africa within the global economy. Among the major debtor nations in Africa are Nigeria, with debt rising from $3.1 billion in 1970 to $13.0 billion in 1985, and Zaire, where debt increased from $3.1 billion to $4.8 billion. In 1985 Sudan had the highest debt-service ratio, 150 per cent, while those for Ivory Coast, Guinea, Togo and Uganda ranged from 40 per cent and 80 per cent (Assiri *et al.* 1990: 118). However, the figures show that the region had the fastest rate of indebtedness in the Third World, with debt stocks trebling over the period.

Latin America and the Caribbean remained the most indebted region in the Third World, though with much variation between countries. For example, the major debtor nations were Brazil accounting for 27.6 per cent, Mexico 25.2 per cent, Argentina 13 per cent, and Chile 4.7 per cent. In the case of Argentina the debt rose from $16.8 billion in 1980 to $50.3 billion in 1987. Comparable figures for the other major debtor nations in Latin America were Brazil (from $5.7 billion to $106.1 billion), Chile ($9.4 billion to $18.0 billion), and Mexico (from $41.3 billion to $96.1 billion) (Assiri *et al.* 1990). This growing indebtedness, together with the deficit on the current account of the balance of payments, has triggered off a constant drain on official reserves. The debt crisis has had an overwhelmingly negative effect on Latin America's institutional development and its quest for sustained growth (Dietz 1987). Of equal signifi-cance is the fact that the debt crisis propelled the region into the hands of the IFIs (International Financial Institutions). As Dietz observed, the region was at the time thrown open into

> restructuring along lines of IMF and neo-classical economic orthodoxy. This means that the role of the state in all its progressive economic and social welfare functions is being sharply reduced, and the economies them-selves have been opened via trade and financial liberalisation to the unimpeded forces of world market competition.
>
> (ibid.: 509)

Third World indebtedness has also been exacerbated by corruption, which meant that state resources were quickly frittered away. This is true of regimes such as those of the generals Suharto, Mobutu and Noriega. These autocrats put in motion structures which ensured that public resources were transformed into private wealth by using bribery, violence and coercion to ward off criti-cisms. In this way, the state is 'privatised', so that which was public yesterday

Table 1.1 Total debt stocks of regional groupings in the Third World for selected years
In US$ million

Year	Sub-Saharan Africa	East Asia and Pacific	Latin America and the Caribbean	North Africa and the Middle East	South Asia
1980	56,825	88,648	242,596	66,394	38,173
1984	84,285	147,575	377,696	92,421	57,684
1986	116,083	186,404	410,946	124,417	80,176
1988	145,916	206,369	428,150	142,745	96,800
1990	173,737	234,685	431,091	141,544	115,351

Source: World Debt Tables 1991–92, Washington, DC: The World Bank, 1991

now belongs to the ruler and the coterie of supporters and sycophants. It was not long before 'kleptocracy' ousted democracy as the new form of governance.[7] The consequences of kleptocracy cannot be measured simply in terms of its immediate impact on the national treasury. As Askin and Collins (1993) have pointed out, it leads to immiseration of the vast majority of the people, the neglect of social and physical infrastructure whilst enriching the ruler, members of his family and ethnic praetorian guards. Kleptocracy destroys public efficiency and rationality as it strengthens patrimonialism and traditionality, by sapping the will of the people to be governed.

For much of the 1980s and part of the 1990s, Africa has been portrayed as an exceptional social formation, one that continuously shows an anathema to modernisation.[8] Central to this theme of the 'new barbarism' is the perception that corruption is an African institution *sui generis*.[9] However, the recent crisis engulfing Asian economies, and particularly the corruption surrounding Indonesia's long-term ruler, has put paid to this extreme form of Afro-pessimism. Asian and Latin American societies have had their share of tyrants, despots and kleptocratic rulers.

We have noted above that it is not interventionism *per se* that has produced the crisis impelling reform and adjustment, but the nature of state intervention in the Third World which has resulted in macro-economic imbalances. Unproductive and irrational expenditure on the military and prestigious projects designed to satisfy elite consumption and bolster support was often the order of the day. Examples in Africa of these types are abundant: the establishment of the steel rolling mill industry in Nigeria which was based not on sound economic principles of 'economies of scale' and market proximity, but on political factors often referred to as 'Federal character', the extravagance and waste of the cement contracts of the 1970s in Nigeria and the frittering away of precious foreign exchange in the 'Abuja contracts'.[10] In the Ivory Coast the Basilica in Yamoussoukro, a state gift to the Roman Catholic Church, constructed at an estimated cost of $100 million, is another example of 'state improvidence'. The wasteful expenditure surrounding Siaka Stevens' decision to host the annual OAU meeting in 1980 at tremendous cost to the nation was an important contributing factor leading to the crisis which engulfed the country so soon after the meeting. The 'financially repressed' citizens of Sierra Leone who had to pay the price for Stevens' egoistic policies quickly coined a maxim: 'OAU for you (Stevens and his party), IOU for me (the citizen)'. In short, government interventions have not increased the 'sophistication and depth of industrialisation' (Stein 1994: 296).

Clearly, not all interventions produce such pathological effects; nor could all the macro-economic problems of sub-Saharan Africa be placed at the door of this peculiar form of 'new economics'. In the absence of a viable or authentic bourgeoisie at the time of independence, African states had to substitute themselves for an investing bourgeoisie, by transforming their own role into a provider of industrial estates, factories and whole industries. In some cases, such as Nkrumah's Ghana and Nyerere's Tanzania, it took the form of joint ventures with foreign investors, total state ownership, and widespread nationalisation (Engberg-Pedersen *et al.* 1996). In the case of Nigeria, various decrees and legislation were enacted to foster an indigenous bourgeoisie (Nafziger 1993). These policies soon became an avenue for rent seeking, fraud and corruption among the professional and political classes (Falola and Ihonvbere 1985).

The Bretton Woods Institutions, the Third World and the neo-liberal consensus

The end of the Second World War, which was accompanied by the rise of the USA as the hegemonic capitalist centre, witnessed the rise of a new international economic order. Both the USA and Britain were instrumental in forging this new order well before the end of hostilities, in order to avoid economic anarchy stemming from competitive devaluation, multiple rates of exchange and other restrictive trade policies (Walters 1994; Ferguson 1988). The plan these two nations set in train was designed to induce international discipline and exchange rate stabilisation. Formal negotiations took place in Bretton Woods, New Hampshire, USA, in July 1944, and this resulted in what became known as the Bretton Woods Institutions (BWIs) or the International Financial Institutions (IFIs). These institutions, consisting of the International Bank for Reconstruction and Development (IBRD), or the World Bank, the International Monetary Fund (IMF), and the International Finance Corporation (IFC), were empowered to oversee international economic relations, and to foster 'the expansion and balanced growth of international trade' (Van Dormael 1978: ix). The participating countries agreed to 'submit to a certain degree of international economic discipline and to delegate certain prerogatives of national sovereignty to a new supranational institution, the International Monetary Fund. The IMF was specifically charged to assure stability of exchange rates, and orderly adjustment if and when this became necessary' (ibid.). This task the Fund continued to perform until September 1971, 'when virtually all the world's currencies were floated after the United States went off the gold standard' (Walters 1994: 10). Despite Third World countries outnumbering the developed nations at the negotiations they 'did not affect to any degree the negotiations or the outcome, (as) the terms of the debate had already been set in the bilateral exchanges between the United States and the United Kingdom' (Ferguson 1988: 26). Furthermore, at the time there was no perception of a clear identity of interests among the developing countries present.

The Bank, on the other hand, was empowered to provide financial resources for countries recovering from the ravages of the Second World War and the economic development of developing nations. The former (reconstruction) task was seen as temporary, thus enabling it to concentrate on its other function of development financing. As Allan Walters has pointed out, 'the Fund is not a fund: it is a bank. And the Bank is not a bank: it is a fund' (1994). As a regulatory body, the IMF determines the rules and regulations governing the activities of member states. Unlike the long-term nature of the Bank's activities, the

Fund was created to provide short-term balance-of-payments assistance, for members who were in difficulties with their external payments. In this way, the Fund would be able 'to avert the imposition of exchange restrictions and other deflationary measures inimical to the attainment of an open trading system and international monetary stability' (Ferguson 1988: 27). Countries which for structural reasons found themselves running short of foreign exchange could obtain short-term aid from the Fund, and the country in question would undertake to abide by or implement certain conditionalities (see below). The Fund also acts as an avenue for channelling assistance to Third World nations, as well as a 'forum for consultation and negotiation of co-operative solutions to international monetary problems' (ibid.: 27). It is important to note that though co-operation between these three institutions is not a new phenomenon, it has become more institutionalised. For example, Kirton (1989) argued that such co-operation was evident in the Fund's discussion with Grenada. Similarly, Hayter and Watson (1985: 113), using an internal published report from the World Bank, have noted that 'the IMF relies on the Bank's judgements on the appropriateness of the country's medium term investment programmes which provides a context for stabilisation measures and is particularly relevant to Extended Fund Facility Programmes (EFF)'.

After 1950 Third World countries embarked on a campaign to obtain financial aid from the IFIs for development and lending to these countries became the mainstays of IBRD's activities (Walters 1994). The period of the 1960s was marked by two developments, which impacted on the IFIs: the intensification of the decolonisation process, and the rise of the European challenge to American hegemony. This new political landscape meant a change in emphasis from reconstruction to development efforts, and the politicisation of the latter (Ferguson 1988). This in turn produced a period of strained relations between the developed capitalist economies and the emerging Third World economies. However, the economic philosophy at the time was premised on widespread state intervention or Keynesian welfarism by both the developed capitalist economies and the Third World.

In Latin America, the Economic Commission for Latin America (ECLA/CEPAL) under the leadership of Raul Prebisch had started questioning the long-term tendency for the terms of trade for Third World (primary) exports to deteriorate *vis-à-vis* manufactured exports. This critique of the division of the international economy in general and the theory of comparative advantage in particular foreshadowed some of the assertions of the dependency theorists, and in this ECLA called, as we saw, for a move away from export-oriented development, towards development from within, i.e. import-substitution development.

By the early 1960s, the Third World had embarked upon a more consistent campaign against the decision-making processes within the international financial system, on the grounds that they were being excluded by the Group of 10 major capitalist nations from major economic decisions; and that the latter sought to sideline the major international financial institutions such as the Fund and Bank and to present them with a *fait accompli*. In order to counter this potential process of marginalisation and collective exclusion, Third World countries formed themselves into a caucus Group of 77 in 1967. This group sought to secure their collective interests by preventing the Fund from being isolated in international monetary relations. Ferguson has pointed out that at this time the Third World saw in the IMF a friendly agency, to which they rendered support. Throughout the 1970s, Third World concern was with the structure of the Fund. However, after 1979 with the accession to power of 'new right' regimes in the United States and Britain whose mission was based on neo-liberal free-market ideas, the concern of the Third World was to oppose what was seen as the harsh conditionality attached to loans from the IFIs. Ferguson described this conditionality as 'by far the most controversial aspect of the Fund's relations with its members' (1988: 198).

Since the Fund can neither be discriminatory in providing loans nor supply funds at penal rates (Walters 1994), it utilises conditionality in order to ration funds. In operation this means 'the economic and financial measures which are needed in particular in order to restore a sustainable external position at the end or toward the end of a Fund programme' (Ferguson 1998: 198). However, Kirton has noted that:

> The nature and details of the EFF programmes often depend on 'who bats for you' within the executive board. If a country is of particular strategic importance and is saying and doing the right things, then certainly it will receive powerful support in the executive board, and usually resulting in softer conditionality terms.
>
> (1989: 142–143)

By the late 1970s, commercial banks' credit had become much more attractive to Third World policy-makers, largely because the banks were awash with petro-dollars and because of the absence of internal control and deflationary effects of the Fund's conditionality. Soon the Fund became an insignificant avenue for balance-of-payments financing for both Asian and Latin American countries, whilst many African countries continued to ignore the Fund by implementing just the minimal conditionality and avoiding the severe adjustment (Walters 1994; Ferguson 1998) which was feared would lead to political instability (Zack-Williams 1997). The end of the 1980s witnessed growing indebtedness for many Third World countries, and for Africa, the decade ended with the continent having been once again hemmed in at the periphery of the world capitalist system.

By the end of the 1980s it was clear that the world had changed fundamentally. There was now a strong tendency towards protectionism, which militated against the prevailing tendency towards a global economy (Leys 1996). Foreign investment had reached new heights to account for between 5 and 10 per cent of total stock in most of the major capitalist economies, and up to a third of world trade was now being undertaken by multinational corporations (ibid.).

Leys has also pointed to the 'internationalisation of capital flow' as one of the striking feature of the post-1979 world economy. He observes:

Instead of merely financing world trade, by the end of the 1980s banks and non-bank financial institutions were dealing in currency exchanges, currency and commodity futures and so-called 'derivatives' of all sorts on a scale that not only dwarfed the conventional transactions needed for trade and investment, but made it impossible for the governments of even large economies to influence the value of their currency by intervening in the currency markets.

(ibid.: 20)

Led by the USA, the world capital markets were systematically deregulated as controls were abandoned. The growing indebtedness of the USA, the world's major trading economy, put added pressure on interest rates, thereby worsening the situation in the Third World.

Apart from the ending of the Bretton Woods system, which regulated international trade, the decade of the 1980s coincided with the rise to power of 'new right' regimes in both the United States and Great Britain, as well as the abandonment of Keynesian economic policy in the Organisation for Economic Co-operation and Development (OECD) (Leys 1996). The subsequent deregulation of capital export meant that certain Third World nations were disadvantaged, since the criteria for attracting capital were now market driven.

Furthermore, these changes impacted on growth rates among the leading industrialised nations. Apart from the new industrialising nations of Asia, growth rates remained low, and this had a negative impact on the terms of trade of exports from the underdeveloped world. Using figures from the *World Development Report*, Leys has noted 'in Sub-Saharan Africa and Latin America the combined effects (of the declining terms of trade and interest rate increases) were estimated to average more than 10 per cent of GDP' (1996: 22). In this context, Glyn and Sutcliffe (1992) have pointed out that 'the share of Africa, Asia and Latin America in world trade is now substantially lower than before 1913' (90).[11] The fall in terms of trade of the vast majority of Third World economies led to declining revenue for their governments and, in an effort to maintain public expenditure, many reacted by increased borrowing, thus setting off the debt trap.

Many African countries began experiencing economic crisis from the early 1970s onwards due largely to an unfavourable international environment and government policies (Nafziger 1993). For non-oil-exporting African states such as Ghana, Sierra Leone, Guinea, Liberia and Uganda, the oil shock had a devastating effect on their balance of payments. Furthermore, the oil price increase with a much broader commodity boom, as well as the need to recycle the petrodollars, transformed the availability of credit in the global banking system (Gordon 1993). African credit-worthiness was enhanced, particularly for the populated oil-exporting countries such as Nigeria, with a desire to promote industrialisation. The net effect was the growing indebtedness of African nations.

These resources were not diverted to productive use, and Africa's export performance remained poor, due to the inability of policy-makers to diversify exports, as primary exports continued to be the mainstay of the continent's foreign exchange earnings. A more serious indictment is the failure of Africa to hold on to its share of world trade. Gordon has observed that during the 1970s Africa's share of non-fuel exports of developing countries fell by more than half, from 19 per cent to 9 per cent. Using World Bank figures he argued that, if Africa had maintained its 1970 share of non-oil primary exports, then its earnings today would be some $9 to $10 billion higher, thus reducing its debt service ratio by half. The situation was exacerbated by drought in the case of the Sahelian countries, and poor food policy which resulted, in the late 1980s, in 'food supplies for minimum levels of subsistence falling below critical levels' (Scott 1992: 201), thus worsening the balance-of-payments deficits, as scarce foreign exchange had to be used to import foodstuffs.

The African economic crisis continued throughout the 1980s, with the annual population growth rate at 3.2 per cent and GDP falling by 1.1 per cent by the end of 1989 (ibid.: 202). Between 1965 and 1973 growth of real GDP per head was 3.6 per cent, but this fell to 0.3 per cent between 1973 and 1980. Whilst the terms of trade continued to deteriorate against African exports, the latter continued to slide, so that between 1980 and 1989, the value of exports declined from $57.8 billion to $39.7 billion, putting additional strain on the balance of payments. The estimated loss to sub-Saharan Africa due to the declining terms of trade from the mid-1980s to the early 1990s has been put at SDR50 billion (Malima 1994: 11). By the beginning of the 1990s, thirty-one countries in sub-Saharan Africa had been characterised by the World Bank as 'debt-distressed' (ibid.: 98), and two-thirds of the world's poorest forty-seven countries are in the region (Logan and Mengisteab 1993: 2). It was not long before the economic crisis was transformed into a crisis of political legitimacy (Olukoshi 1994b; Lander 1996) with the authority of the governing classes being challenged by social movements ready to contest state hegemony. In an attempt to deal with the deepening crisis of budget deficits, falling revenue, chronic balance-of-payments problems, gradual decay of infrastructures as well as social welfare facilities, many of these countries experimented with IMF and World Bank structural adjustment programmes. Some countries, such as Nigeria (Oluyemi-Kusa 1994) and Zimbabwe (Seshamani 1994: 114), have instituted home-grown structural adjustment programmes; whilst others such as Sierra Leone followed a 'shadow programme' (Zack-Williams 1992 and 1993) for a period, following the unilateral abrogation of the agreement by the IMF due to failure to keep up with debt repayment. However, as Wapenhans has observed, 'adjustment without external financial assistance was in most cases likely to cause unsustainable social pain' (Wapenhans 1994: 37). The vast majority of countries in the region have followed IFI-sponsored structural adjustment programmes to varying degrees.

Nonetheless, the shift from the African version of new economics (Preobrazhensky 1965) is not the product of simply the failure of the developmental efforts in Africa alone. As we have seen, the period 1973–1983 witnessed in the developed capitalist societies 'a shift in economic policies from an interventionist stance, which permits and sometimes encourages state intervention in the economy, towards a neo-liberal position which aims to minimise it, letting the market allocate resources wherever possible' (Engberg-Pedersen *et al.* 1996). This shift heralded the triumph of the neo-conservatives in the developed capitalist societies, particularly in Britain and the United States. Their economic priority was premised on the need to reduce inflation, which had been induced by two oil price rises in the 1970s, by embarking on deflationary policies (Toye 1994). The latter triggered off severe crises of indebtedness in the underdeveloped world, which had been building up throughout the 1970s (ibid.).

Conclusion

In the preceding pages we have drawn attention to the factors which impelled crisis-ridden Third World states to adopt IFI-sponsored structural adjustment programmes. In particular, we have looked at how Third World countries became 'hemmed in' on the long road to structural adjustment. In looking at this process, we have drawn attention to the fact that SAPs are the result of the failure of post-colonial developmental efforts and that SAPs themselves are not new nor are the conditions imposed by the international financial institutions, though the latter have tended to change over the years. We also noted that the underpinning logic of neo-liberalism runs counter to the statist approach which defined the post-colonial development efforts of most Third World states.

To conclude this chapter on the long road to adjustment we consider the various ways in which the neo-liberal agenda (of which SAPs are but one, albeit important, element) has been linked with the rapid processes of globalisation that appear to characterise the contemporary international scene. Whilst the study of SAPs has produced a flood of analyses, it has only been a trickle compared to the amount that has been written on the intensification of global economic linkages and other 'indicators' of political and cultural globalisation during the past decades, particularly since the end of the Cold War (Waters 1995). Much of this literature, however, presents globalisation as a benign (or disembodied) and apolitical process of the growing integration of the distinct regions of the globe (Giddens 1993).

Dissonant voices on globalisation

Recent theorisation on neo-liberalism has been based upon the premise that an effective understanding rests upon considering its relationship to globalisation – an identifiable and agreed-upon transformation in the way in which the world operates. Thus, neo-liberal apologists for market-led reform talk, not only of the theoretical merits of market liberalisation, but also of the inescapable logic of economic integration, of responding to opportunities and not getting left behind in the wake of globalisation. Similarly, most of the more critical views of the relationship between neo-liberalism and globalisation also assume that some fundamental transformations have occurred in recent years and whilst their interpretations stress the inequalities of globalisation they share the neo-liberal belief in the all-encompassing nature of the transformations and the lack of opportunities for individual nations to pursue alternative strategies. For example, in a speech delivered in 1995, Tony Blair stated that:

> What is called globalization is changing the nature of the nation-state as power becomes more diffuse and borders more porous. Technological change is already reducing the power and capacity of government to control its domestic economy free from external influence.
>
> (cited in Dicken *et al.* 1997: 159)

Some have depicted such perspectives as overly pessimistic, debilitating to the pursuit of alternative perspectives and conceding too much to the neo-liberal position. Such authors generally concede that there have been profound changes in the international economy and individual societies over recent years and share the view that such changes have been directed by a range of particular interests but have refused to consider these changes as constituting some new 'global' era with its associated impact upon national economic decision-making, etc. Positions of this type have led to fierce exchanges in the literature between those positively committed to the 'globalisation' thesis (perhaps most strongly expressed in the work of Kenichi Ohmae – see Ohmae 1990), those who, whilst critical of its effects, believe that a profound process of globalisation is occurring (Hoogvelt 1997), and those who see it as overblown and unhelpful (see Hirst and Thompson 1996).

Interpreting the rise of neo-liberalism and its significance

Advocates of the market-led reforms that have underlain the globalisation process present the changing policy-focus of recent years as simply the application of 'sensible' economic management. With the collapse of the socialist alternative in Eastern Europe and the crisis within more interventionist Keynesian approaches to economic management and the West European welfare state, something of a consensus (often termed the Washington consensus) emerged within mainstream economics that the market-led strategies of neo-liberalism represent the most appropriate form of economic management. More significantly, however, it is also suggested that neo-liberal economic policies are particularly suited to the new global era; that the embracing of liberalisation and market-led reforms is the only way to take advantage of the opportunities of the global market place and provide opportunities for all countries to improve their economic efficiency and, by inference, the living standards of their citizens. In this sense, defenders of neo-liberalism tend to see such policy frameworks as an appropriate response to the phenomena of the globalisation of economic activity (through embracing and encouraging it) rather than as the main instigator of it.

It is clear, however, that globalisation is not something that is simply happening 'out there' that must be responded to. The integration process has been shaped by the decisions taken by national governments through international negotiation as well as through the actions of transnational capital. It continues to be actively promoted in many quarters; through the pursuit of free trade in the World Trade Organisation (WTO) and international trade negotiations, through regional free trade agreements such as the NAFTA, through international agreements on the conditions for investment (MAI), and so on. In this sense, rather than being a reaction to the globalisation process, neo-liberalism might be better conceived as the, often unspoken, ideology that has actively promoted, and to a certain degree created, globalisation. As Marshall (1996: 887) puts it:

> The changes in the techno-industrial base of the world economy would have been ineffectual without the rise of neo-liberalism and financial liberalisation. The reification of the 'market' as a neutral and natural institution, apolitical and ahistorical, has now become common in academic and policy circles.

None the less, despite the obvious connections between globalisation and the neo-liberal political project, an analysis of the political agenda that underlies the drive towards the continued intensification of global linkages (and the differential spatial impacts that have ensued) is strangely missing from many treatments of the globalisation phenomena. An examination of the origins, role and dynamics of the neo-liberal ideology and its relationship to the global processes it has overseen are, therefore, important in gaining an appropriate understanding of the changing nature of the global economic system. As expressed by Overbeek and Van der Pijl (1993: 2), 'How can we account for such a process? [the consolidation of neo-liberal hegemony]. Is it the outcome of a battle of ideas, or is it the product of the concrete agency of social forces? If so, does that then mean that social forces are capable of redefining the coordinates of what is considered "normal"?'

An examination of the social forces underlying the neo-liberal agenda leads some to see neo-liberal reforms as part of the gradual emergence of a new global economic order designed to deal with the crisis in the post-war capitalism that occurred at the beginning of the 1970s. To them, SAPs and the other expressions of the neo-liberal agenda are seen as reforms that, despite the free market rhetoric, 'regulate' capital accumulation at a world level to the benefit of the interests of dominant economic and political elites. To some, this is best conceived as part of a global imperialistic strategy designed to keep developing countries in the position where the West wants them. Assessments of this kind are, in essence, a rejuvenated form of dependency theory. As suggested by Duncan Green (1995: 128):

it is hardly surprising if many in the Third World tend towards conspiracy theories which see the free trade crusade as part of a deliberate policy of 'global rollback' by the US and other industrialized powers, using the power of the IMF, World Bank, GATT free trade agreements and US trade legislation to *prevent* industrialization in the South and to force the Asian economies to open up to US investment and trade. Each tightens the strait-jacket which has been placed around Latin America's economic future, forcing it along the neo-liberal road, while increasingly depriving its people of the right to choose a different destiny, should neo-liberalism turn out not to be the nirvana promised from Washington and Geneva.

Others have developed a more theoretically sophisticated account of the relationship between neo-liberalism and global restructuring that makes more of the relations between classes and the forms of regulation of the global economy rather than the more simplistic model of North–South relations. Overbeek and Van der Pijl, for example, cast their discussion of the rise to hegemony of neo-liberalism in terms of what they call 'comprehensive concepts of control'. These concepts relate to specific historical configurations of classes and class fractions and particular configurations of capital that have occurred in the evolution of the global economic system (Overbeek and Van der Pijl 1993: 3). To them, neo-liberalism emerged out of the crisis of the late 1970s when one configuration of control (corporate liberalism) was no longer able to function effectively and was gradually replaced by another, neo-liberalism, which was able to restructure 'both spatial and technical aspects of production and the social relations of production, in order to adjust production to consumption, and restore profitability by raising the rate of exploitation and the mass of surplus value' (Overbeek and Van der Pijl 1993: 18).

Whilst certainly important, the generalised treatments of neo-liberalism and globalisation considered thus far often miss the intricacies of how the neo-liberal agenda has been received and implemented in the countries of the South, who are often presented as helpless in the face of relentless global changes and powerful international interests. Whilst undoubtedly having some validity, such characterisations tend to miss the self-fulfilling nature of the more pessimistic international conspiracy scenarios and understate the regional differences in both the globalisation process itself and the way in which the neo-liberal discourse has entered into specific social formations and political cultures. As again expressed by Duncan Green (1995: 176–177):

> Geography (for example proximity to the US), the state of the world economy at the time of adjustment, the size of the domestic market, the availability of natural resources or a skilled workforce, the capacity of the civil service, and the prior existence (or absence) of a dynamic local private sector, besides the political preferences of different presidents, all influence the way neo-liberal ideas are put into practice.

In the rest of the book we intend to examine these contentions. As we asserted in the introduction we believe that while general processes may underlie adjustment experiences it is inaccurate and politically debilitating to reify these processes because doing so precludes the formulation of locally- and nationally-specific responses. It is precisely these responses, actual or potential, with which we end the book.

Notes

1 As Callaghy and Ravenhill observed, 'By "hemmed in" we mean a situation in which the viable policy alternatives, and the capacities and resources needed to implement them, available to African Governments are severely constrained as a consequence of volatile politics, weak states, weak markets, debt problems, and an unfavourable international environment' (1993: 2).

2 This refers to clientelist networks used to build support via rent extraction, expansion of the state sector and state support through subsidies for the vociferous urban dwellers. See Callaghy 1990.

3 *African Development Lessons from Asia*, Winrock: International Institute for Agricultural Development, 1991.

4 See C. Young, *Ideology and Development in Africa*, New Haven: Yale University Press, 1982, for this taxonomy.

5 In 1970, the government of Sierra Leone took a controlling share holding in the privately owned mining concern, the Sierra Leone Selection Trust, to form the National Diamond Mining Company.

6 Figures from *World Debt Tables 1991–92: External Debts of Developing Countries*, Vol. 1, Washington, DC: World Bank, 1991.

7 For the case of Zaire see Askin and Collins 1993: 72–85; for the case of Sierra Leone see Zack-Williams 1990: 22–33.

8 See, for example, the work of R.D. Kaplan, 'The Coming Anarchy: How Scarcity, Crime, Overpopulation, and Disease are Rapidly Destroying the Social Fabric of the Planet', *Atlantic Monthly*, February 1994: 44–77; and that of Fukuyama 1992.

9 For a sustained critique of this position see Richards 1996.

10 The 'Abuja contract' refers to the vast amount of money wasted in offering contracts to people who had no intention of fulfilling their side of an agreement. To go to Abuja for contract was a feature of business life in Nigeria in the 1970s and early 1980s.

11 Glyn and Sutcliffe 1992.

2 What is structural adjustment?

Bob Milward

Introduction

The consequences of structural adjustment lending are far reaching and global in their nature. Hence, anyone with an interest in development, whether from an economic, sociological, political or environmental point of view, must understand the implications of structural adjustment for those economies that must resort to this type of assistance. In what follows we will examine the development of structural adjustment, outline its major elements and discuss the theoretical underpinnings of the conditionalities that are attached to the lending. This analysis will therefore enable an understanding of the context within which the subsequent chapters develop the role, the practice and the outcomes of structural adjustment over the past two decades.

The development of structural adjustment

As the previous chapter showed, the introduction of structural adjustment loans coincided with the appearance in the global economy of a large group of underdeveloped nations who were experiencing rapidly deteriorating economic indicators and serious macro-economic problems. These problems were identified with a series of external, global circumstances that had a deleterious effect in all areas of the global economy. However, it was recognised that deficiencies in national policy-making processes in underdeveloped economies and structural weaknesses in their economies were a significant, contributing factor to their worsening economic performance. To address this state of affairs, the institutions of the Bretton Woods agreement, the World Bank and the International Monetary Fund (IMF), initiated a reorientation of economic policies, particularly concerning stabilisation policies and structural adjustment programmes.

As we have pointed out, whilst a number of these countries had structural adjustment programmes in place by the end of the 1970s, the list of adjusting nations grew significantly in the 1980s, and by the end of the decade only those few countries where internal conflicts made it difficult for organised activities to take place were exempt, 'and even for them, the requirement to undertake specified measures in order to qualify for financial support is almost an obligation'

(Malima 1994: 10). The growth of SAPs in the 1980s is symptomatic of the fact that, in the period 1980–82, average value per year of adjustment loans was $190 million; rising to $468 million in the period 1983–1985; $1,241 million for 1986–1988; finally for the period 1980–91, it was $10,025 million (Sparr 1994).

Essentially, structural adjustment is the process by which the IMF and the World Bank base their lending to underdeveloped economies on certain conditions, pre-determined by these institutions. The pre-conditions concern the drafting and implementation of economic policies that are acceptable to the institutions themselves. The prevalence of SAPs as a curative measure for the macro-economic imbalances within Third World economies marks the triumph of mono-economics over structuralism (Toye 1994: 22). Structural adjustment is a 'generic term to describe a conscious change in the fundamental nature of economic relationships within society' (Sparr 1994: 1). It is designed to enable the adjusting country to change the structure of its economy in order to meet its long-term needs of efficient utilisation of factors of production to ensure sustained growth (Woodward 1992). More specifically, it refers to the process whereby the economies of the Third World are being reshaped to be more market oriented.

The rationale for such policies has its modern roots in the 1970s, when two major factors influenced the long-term economic policies of most of the developed nations and financial institutions of the world. These factors were, first, the two enormous oil price increases imposed by the Organisation of Petroleum-Exporting Countries (OPEC) in 1973, when the price of oil quadrupled, and a further tripling of the price in 1979/80. The outcome was a raw commodity price boom, which was to benefit many developing nations in the short term, and also presented the opportunity to international financial institutions, who were in receipt of large increases of petrodollars, to increase their lending to developing economies. Much of this lending took place without a great deal of thought as to its relevance, or an understanding of the economic problems that were facing the underdeveloped nations in the longer term. Many of these economies were internally very weak, attempting to develop at an accelerated rate, with an over-reliance on their primary product exports in a highly unstable world market. The oil shocks served to bring to the surface the weak position of many underdeveloped economies, relative to the developed countries and, as such, when the second oil shock sent the global economy into a deep recession, the major impact was in terms of a large decrease in the demand for raw materials, as the industrial nations reduced their imports. The second, but interrelated, factor was the rise in real interest rates, from an average of 1.3 per cent between 1973 and 1980, to an average of 5.9 per cent between 1980 and 1986 (Toye 1994: 20). This increase was due to the introduction of neo-liberal economic policies, commonly described as being 'monetarist', in the major Western industrial nations of Europe and North America, partly as a reaction to the inflationary pressures that emerged in the aftermath of the oil price shocks.

These policies included economic austerity measures to squeeze out inflation, causing large increases in unemployment and the depression of demand. As such, these deflationary measures served to undermine export growth in the developing countries in the 1980s. The position of the industrial nations at the onset of the recession was relatively strong, whilst many of the underdeveloped countries were plunged into financial crises as the terms of trade went further against them, their balance-of-payments positions reached crisis point and the interest rates on loans received from the international financial institutions rose so sharply that an increasing amount of their foreign exchange earnings, already reducing in supply due to the recession, went towards the servicing of the nations' debt repayments. The financial institutions came to realise that the prospects for recovery of the full principal and the interest payments were becoming increasingly unlikely due to the deflation of the industrial economies and their lack of supervision of the investment strategies of the borrowing countries, where much of the lending had been used to fund low-productivity projects.

The role of the international financial institutions

The agenda of the IMF was for conditionality lending, but gradually under the new economic climate of the 1980s, the World Bank moved closer to the IMF model (Bird 1994). This was as a result of a growing disillusionment with the current methods of funding and advising nations on how to implement economic policy in order to achieve economic growth, greater output and effi- ciency. The advice was given as 'policy dialogue', and project funding was regulated in respect to a country's response to the advice that was given. Several flaws in this system of funding were identified over time. Policy dialogue could well be ineffective if there were many projects in the pipeline and, therefore, the World Bank found that both project financing and unproductive dialogue could run alongside each other. There was also the so-called 'fungibility' problem, whereby the question was raised as to whether a borrowing nation could have initiated the project without recourse to World Bank project financing. That is, the project is funded by external financial institutions, releasing domestic resources for expansionary government expenditure. In sub-Saharan Africa, the growing problem was the financial constraints of the world market in recession and the fact that the governments in these countries were increasingly unable to provide and sustain the economic environment within which funded projects could succeed. By the late 1970s, the number of funded projects in sub-Saharan Africa had reduced significantly and, in the eyes of the World Bank and the IMF, most of the countries did not have acceptable macro-economic strategies.

With the rise of the neo-classical, monetarist ideology in the 1970s, Ernest Stern, the head of the operational arm of the World Bank, felt that in his opinion, structural adjustment lending was a much more efficient and effective method of achieving policy changes in the borrowing countries than policy dialogue (Mosley *et al.* 1991: 33). Stern believed that once these countries took note of the policy changes that had been advised by the World Bank, and acted upon them, efficiency and greater output would be achieved, and that through the free-market mechanism, poverty would take care of itself through the 'trickle-down' effect. This reorientation of policy designed to encourage economic development in underdeveloped economies was endorsed by Robert McNamara, then the president of the World Bank, at an UNCTAD (United Nations Conference on Trade and Development) meeting in Manila in April 1979. McNamara stated that the way to achieve economic development was to put forward assistance to countries who would be willing to align their economic policies to those suggested by the World Bank, and in doing so, create the conditions that would be conducive to greater and faster development. Initially, the board of the World Bank had some reservations as to the likely success of the new strategy, but approved the World Bank's management decision. The major reservations centred upon the reluctance that countries had shown in the past to the implementation of changes suggested to them in the policy dialogue process, due to the recognition by governments in underdeveloped economies that the policy proposals could damage their domestic position. As such, insisting on policy reform through structural adjustment lending would, it was thought, inevitably enmesh the World Bank in a country's internal politics, removing its supposed political neutrality.

In 1986, the IMF took the decision to provide assistance to low-income, developing countries facing protracted balance-of-payments difficulties. The assistance was dependent upon agreement by the government of the country concerned to undertake medium-term structural adjustment programmes, prescribed by the Fund, with the intention of fostering economic growth, and of strengthening the balance-of-payments position. Countries that were eligible for assistance were required to prepare a three-year adjustment programme set out in a 'policy framework paper', with the approved loans to be over a ten-year period at a special initial interest rate of 0.5 per cent. However, the Fund is also conditioned by the beliefs and philosophy of neo-classical economic theory, which favours the free movement of capital. As such, a balance-of-payments deficit is seen as being caused by price uncompetitiveness and excess demand. The 'cure', therefore, is an admixture of devaluation and demand contraction, and a restoration of the price mechanism for the allocation of scarce resources.

Types of financing available from the BWIs

Those supported by the IMF are:

- Stand-by arrangements (SBAs): loans at commercial rates which provide support for macro-economic adjustment programmes, usually lasting 1–2 years.
- Extended arrangements (usually termed extended fund facilities – EFFs): loans lasting three years and given to those countries showing a strong commitment to adjustment.

(continued on p. 28)

- The Structural Adjustment Facility (SAF) and Enhanced Structural Adjustment Facility (ESAF): loans at concessional rates available only to low-income countries. Usually lasting three years, they cover both structural and macro-economic adjustments and are drawn up in collaboration with the World Bank.

Those supported by the World Bank are:

- Structural adjustment loans (SALs): loans to support structural adjustment programmes affecting the whole economy.
- Sectoral adjustment loans (SecALs): loans to effect adjustments in specific sectors (e.g. health, energy).
- Hybrid loans: loans which include elements of SecALs and traditional project loans.

(Woodward 1993)

The neo-classical paradigm

> The World Bank model is based very firmly on the orthodox neoclassical view of economics – particularly the efficiency of free markets and private producers, and the benefits of international trade and competition.
>
> (Woodward 1993: 4)

To understand more fully the motives that lie behind these components of the structural adjustment packages, we need to examine the neo-classical paradigm in greater detail. The Bretton Woods Institutions justify their interventions using a mixture of these theories and a range of discursive strategies which 'prove' that benefits will accrue from the application of adjustment reforms (Mohan 1994). The reforms that are put forward can be categorised as external reform and internal reform, but with both based firmly in the neo-classical scheme.

The essential concept is that of equilibrium, and free markets and free trade are employed as the best methods of achieving an equilibrium both internally and externally. The basic model is that of supply and demand and this is illustrated in figure 2.1. Here the supply curve (S) slopes upwards from the left, showing that the higher the price, the more output that will be produced, the higher the price the greater the incentive to produce. On the other hand, the demand curve (D) slopes downwards, from the left, indicating that the lower the price, the higher will be the demand for output from consumers. Hence, the two sides of the model would appear to be incompatible, the firms, represented by the supply curve, wish for higher prices and consumers, represented by the demand curve, wish for lower prices. Hence, if the price charged by the firms is too high, then there will be a surplus of output (Q^I–Q^{II}) which remains unsold because at the price P^I, output will be Q^I (point A), but demand for the product will only be Q^{II} (point D). Conversely, if the price is set too low, then some consumers who wish to buy the product will be disappointed due to a lack of supply. At price P^{II}, the level of output that consumers would wish to buy is Q^I, but the supply of output is only Q^{II}. Hence, the only point at which both producers and consumers are satisfied is the output level Q^e at price P^e, giving us the equilibrium point, E, where there is no surplus of production and no unsatisfied consumers who are willing to pay this price. It follows therefore that this equilibrium point is the point of maximum efficiency. This basic model is then used to show how free markets, that is markets that are allowed to find the equilibrium point through the interaction of supply and demand, will produce the most efficient outcome in aggregate for the economy as a whole.

Internal policy reforms

The same model can be used in the case of the market for labour. In figure 2.2, a wage above the equilibrium wage will produce unemployment because at the wage W^I the demand for labour will be lower than at the equilibrium wage, W^e. The policy-makers use this to argue that the state-controlled labour market keeps wages artificially high, thereby creating unnecessary unemployment. By cutting wages, it is argued, production will shift to more labour-intensive producers and subsequently reduce unemployment.

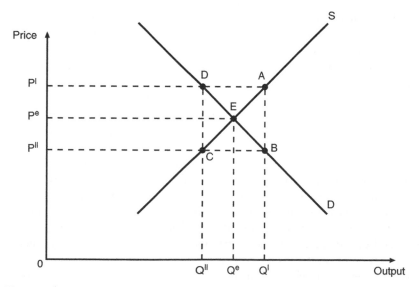

Figure 2.1   Market equilibrium

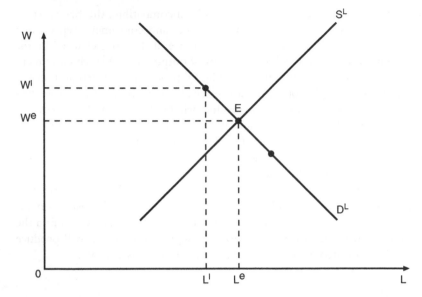

Figure 2.2   Equilibrium in the labour market

The theme continues into the structure of industry in the economy. Here, the contrast is between perfect competition and monopoly. Figure 2.3(a) is the case of perfect competition where a large number of firms all produce the same product, have perfect information and must take the price that is given to them by the equilibrium of supply and demand in the industry as a whole. At this equilibrium price, each firm will produce the level of output that is most efficient. That is, at the lowest point on its average cost curve (AC). In addition, the firm will be maximising its profits where marginal cost (MC) equals marginal revenue (MR) and this point coincides with the efficiency level of output. Hence, each firm is maximising its profits at an efficient level of output and the industry as a whole is in equilibrium. Should the firms in the industry make higher profits, then because there is perfect information and no barriers to entry into the industry, other firms will be attracted to the industry, supply will increase and the equilibrium price will fall, such that excess profits are removed and the industry returns to equilibrium.

In contrast, figure 2.3(b) illustrates the case of monopoly where there is only one firm in the industry and that firm has the ability to either set its price or its level of output. If the firm is assumed to be profit maximising, then it will produce where its marginal cost equals marginal revenue; in the diagram this will result in an output of Q at a price P. However, we can see that this is not an output level of the greatest efficiency because the monopolist is not producing at the lowest point on its average cost curve. In other words, the monopolist at its profit-maximising level of production is restricting output because it could produce more at a lower price, and is making excess profits. This situation will

(a) Perfect competition

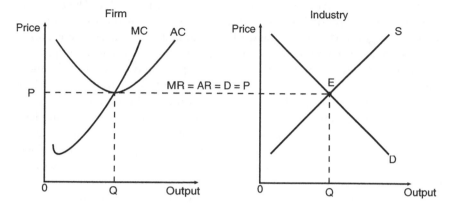

(b) Monopoly

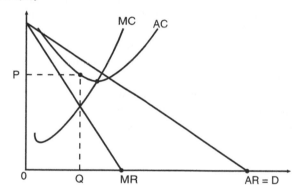

Figure 2.3 Structure of production

persist because there are assumed to be barriers to entry into the industry, and hence, it is concluded that this is an inefficient structure for the industry. It follows, therefore, that measures to increase efficiency in an industry must, according to this theory, involve the reduction of barriers to entry and the promotion of competition, giving the rationale for a privatisation and liberalisation programme in the economy. Hence, structural adjustment programmes contain measures to disaggregate the supply side of the market through denationalisation and the removal of state subsidies with the objective of increasing competition and, according to the theory, raising the efficiency of the internal economy.

External policy reforms

In terms of the external economy, the neo-classical model of trade is concerned with free trade, that is trade without restriction, using the model of comparative advantage. The advantages of free trade are shown in figure 2.4. The line PM^{II} is the production possibility curve of an economy, given the two outputs of good X and good M. Hence, using all of its available resources, the economy can produce a certain amount of each good. The amounts that it will produce are given by the intersection of the production possibility curve with the indifference curve I^{I}. Here, utility will be maximised and the output of good M is M^{I}, and the output of good X is X^{I}. Now, consider the situation where the country concentrates production on good X; it can now produce a higher level of output and export its excess production, using the proceeds to import good M. The terms of trade are represented by a negative slope, with steeper slopes showing more favourable terms of trade (that is, higher relative prices for exports). The economy can trade along any line with the slope that represents the terms of trade. Here it can trade along PB. Since the economy is open to trade, output and consumption do not have to coincide as they do in a closed economy. Thus, the optimal point for production is P, as this allows the economy to attain higher consumption levels by trading up to point PB than producing at any other point. Consumers will face a price equal to the terms of trade and utility is maximised on the higher indifference curve (I^{II}). Because the country is no longer producing good M, its consumption level of this good must be imported and these imports are paid for by the export of good X equal to the amount TP, with the amount 0T being consumed in the domestic economy. The outcome is mutually advantageous, with each country specialising its production in the most efficient outputs.

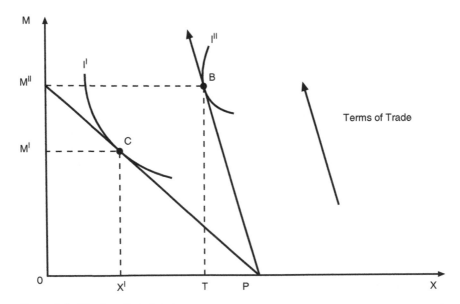

Figure 2.4   The benefits of trade

In a widely quoted study for the World Bank, Balassa (1981) produces comparative data which show that those countries which were least affected by the oil shocks of the 1970s were those which were 'externally-oriented'. He states that 'The findings of the investigation point to the advantages of outward-oriented policies for export performance and for economic growth in the face of external shocks' (1981: 5). Similarly, Bruno (1988) identifies 'a very high initial degree of inflexibility and "closedness" as the starting point for reform' (223). In effect, the analyses provided the blueprint for adjustment programmes as its author proclaims that 'internal shocks which may find their origin in inappropriate policies ... adversely affect economic growth and the balance of payments, requiring the application of structural adjustment policies' (Balassa 1981: 1). Governments should remove themselves from setting market prices while 'growth objectives will need to be given greater weight as compared to income distributional objectives' (2). In this short document we see all the elements of the adjustment package coming into place as social welfare provision should be cut as it 'encourages consumption', which these governments can ill afford.

This then provides the theoretical underpinning for the policy prescriptions of the neo-classical paradigm employed by the Bretton Woods Institutions in the structural adjustment programmes. Free markets and free trade are supposed to provide internal and external equilibrium and the forces of competition then produce increasing efficiency in the use of scarce resources, providing the economic growth that is vital for economic development. It is therefore argued that governments in underdeveloped economies have often intervened to a far greater degree than neo-classical principles would suggest, resulting in a lack of economic growth and the absence of economic development. The remedy must therefore be to free prices from government intervention, in terms of regulation and subsidy, to privatise state-owned firms, to reduce state expenditure in favour of private investment and provision, and to remove artificial barriers to trade, such as tariffs and quotas.

The policy options

Given this neo-classical paradigm, the reforms that are required under a structural adjustment package have the intention of freeing prices and markets from the influence of the state. As such, the main components of structural adjustment encompass five major areas of reform:

1 To allow markets to work by allowing the free market to determine prices.
2 To reduce the state control on prices, such that prices may be set by scarcity values.
3 The divestiture of resources held by the state, into the private sector.
4 The reduction of the state budget as far as is possible.
5 A reform of state institutions to reorientate the role of the bureaucracy towards the facilitation of the private sector.

Within this, each of the components contains specific reforms to achieve the desired outcome. In the next part we will evaluate the impacts of these reform areas. First, to allow markets to operate freely and for prices to reflect scarcity values, trade reform has to be undertaken to eliminate import controls, to reduce tariff levels, to provide a uniform tariff structure and to reduce existing restrictions on foreign investment (Dornbusch 1988). Second, the exchange rate should be adjusted to provide for the profitability of export industries and interest rates set at levels that are above the rate of inflation to reduce, and then eliminate, the excess demand for credit (Little *et al.* 1993).

Third, any regulations that impinge upon the labour market, such as minimum wage controls, must be removed, as must controls over prices in other markets. In particular, farm product prices should be free to find their own level, leading to increased productivity in the agricultural sector, which in turn leads to higher investment in the sector. Fourth, the divestiture of resources held by the state entails a privatisation programme which transfers the assets held by the state to private investors. In addition, government-provided services should be contracted out to the private sector and any state expenditure that remains should be concerned with the provision of an economic environment within which the private sector can operate more efficiently. For example, expenditure on the infrastructure, on health and on education. However, the fourth area of reform, reduction of the state budget, means that state expenditure must become more efficient through rationalisation of the workforce and greater accountability of government agencies and devolved budget to create greater efficiency.

In practice, these five areas are split into two phases. SAPs contain two major components: first, stabilisation and, second, demands for market liberalisation and public sector reforms (or adjustment). Negotiations between the IMF and a financially troubled nation will result in stabilisation loans, for between twelve and eighteen months, or Extended Fund Facility lasting up to three years, designed to finance austerity plans. This stage could be preceded, as in the case of Benin in 1984, by a Country Economic Memorandum (CEM), serving as the 'principal instrument for its economic policy dialogue with the authorities' (Westebbe 1994: 85), and full agreement between the IFIs and the Beninois authorities was not reached until 1989 when the latter dropped its opposition to SAPs by embracing a 'radical programme of structural reform that represented a fundamental shift in development strategy' (ibid.: 88). This was quickly followed in May 1989 by a structural adjustment loan (SAL) of $45 million provided by the World Bank; by an IMF Structural Adjustment Facility (SAF), as well as significant inflow of funds from the EU, and Switzerland (ibid.). The aim here is to inject sharp, short, shock treatment to the economy in order 'to reduce internal and external dis-equilibria and promote growth'. For example, the Ugandan ERP of 1987 included currency reform and the introduction of a new currency (Henstridge 1994: 206).

By demanding cuts in government expenditure, reduction in the size of the bureaucracy, ending of subsidies, devaluation, deregulation of exchange rates and price control, ending of state monopoly in exports (such as the marketing boards in West Africa), as well as in the distribution of essential commodities such as rice, stabilisation mechanism quickly leads into economic recession, with mass unemployment and hyper-inflation. To cushion the effects of stabilisation once a government has these policies on board, funds are made available from which it can draw periodically as long as certain economic criteria are met including payment of arrears to the Fund. Once the economy is stabilised, adjustment moves on to the next stage, that is more long-term corrections via structural adjustment loans. In many sub-Saharan African states this stage is never arrived at before the Fund abrogates the agreement because of culpable slippage by the adjusting government; or before austerity produces 'IMF Riots' (Walton 1995) leading to the abandonment of adjustment.

Critique of the neo-classical model

However, there are serious criticisms of the neo-classical model to be made, particularly in terms of its underlying assumptions. We will see in subsequent chapters in part II how these contradictions unwind in the operation of 'actual' SAPs.

The essential assumption in the supply and demand model of figure 2.1 is that of consumer sovereignty, where demands are initiated by consumers and firms play a passive role in terms of only reacting to the demands of consumers. This demand is triggered only by what is termed 'effective demand', that is through the economic votes of consumers using money. In the real world, however, there is a lack of consumer sovereignty due to consumer ignorance and firms will use this to their advantage by providing information about their products through advertising, pushing those products which enable the greatest amount of profit to be made. In addition, there does not exist an equality of economic votes and as such, the demands of the rich are more likely to be satisfied than the demands of the poor. If there is therefore an unequal distribution of income and wealth in the economy, then this will be likely to be reinforced by market forces.

In the labour market of figure 2.2, there is an implicit assumption that the two sides of the market, the producers who demand labour, and the workers who supply their labour, have equal power in the market. This again is shown to be a situation that does not exist in the real world. Unless workers combine to present a united front, the producers have far more power in the bargaining relationship of the labour market (see chapters 4 and 5). The industrial structure of figure 2.3 has unrealistic assumptions of the perfect knowledge of buyers and sellers and the absence of barriers to entry. In addition, it could be argued that in certain circumstances the existence of monopoly would be preferable to that of perfect competition. For example, in the case of so-called 'natural monopoly', where it would be inefficient for more than one unit of production to produce a particular good. Examples of natural monopoly are to be found in the electricity grid, where competition would produce a duplication of electricity cables and pylons, railway tracks, water pipes and roads. All would entail a waste of resources if competition were to be introduced. It can also be argued that a vital component in economic growth is the introduction of new techniques of production which increases productivity. In the case of perfect competition, the profits that are made are sufficient only for the firm to continue into the next period and investment from retained profits could not take place. Even if an individual firm wished to improve its efficiency by introducing new technology, the incentive to do so does not exist. This is because there is assumed to be perfect knowledge and, therefore, all firms will have access to the same knowledge and in the long run all firms can only make normal profits. Hence, perfect competition would produce a static economy. On the other hand, the excess profits of the monopoly firm could be used to invest in research and development with the incentive being the higher levels of production that could be achieved and the preservation of the barriers to entry. There is also the benefits of economies of scale to be reaped in a monopoly situation. This would produce a dynamic sector that is continuously increasing productivity and efficiency.

In the model of international trade, there are two assumptions in particular that should be criticised. The first of these is the underlying assumption of full employment of all factors of production, which is clearly the least realistic in underdeveloped economies. Second is the assumption that economies have equal influence in their trade with each other. This does not exist in the contemporary world where, in fact, there is an unequal exchange between the primary product producing underdeveloped economies and the products of the advanced industrial economies. The outcome is that free trade produces an increasing gap between the national income of the poor and rich nations.

All of these problems with the neo-classical model can be termed as market failure, and whilst neo-classical economists tend to see market failure as the exception to the rule, an examination of the model tends to suggest that it is the model itself that is the exception. This is the crucial bone of contention between the different sides of the argument, with the neo-classicals arguing that market failure comes about through governments intervening in markets and distorting the outcomes and others who suggest that governments must intervene to prevent the anti-social outcomes of a totally unrestrained free market.

Market failure and state intervention

There are market failures that both the neo-classicals and their critics would agree require state intervention, and these are the cases of externalities, public goods and merit/de-merit goods. Externalities, or the external costs of production, occur where the activities of a firm produce a cost that has to be borne by others. For example, pollution or the overuse of natural resources such as forests. Hence, firms do not pay the full costs of their production and produce more output than is optimal for society as a whole. In the same way, firms may underinvest in activities where the benefits are realised outside the firm, such as the training of staff in non-firm specific skills which workers can then use to bargain for higher wages in other firms which do not train their workers, saving on training costs and therefore able to pay higher wages. Merit/de-merit goods are those goods that, if left to the market, would be underproduced or overproduced. Cultural amenity could be considered to be a merit good in that access for the whole of the population could lead to a more enlightened, and therefore more harmonious, society. De-merit goods include drugs, firearms and alcohol, which if left to the free market would be available in large quantities and at low prices, producing large problems for society. Public goods contain elements of both externalities and merit goods but contain the concept of the 'free-rider'. For example, society may generally agree that national defence is a good thing, but individuals would not make a voluntary payment for its provision in the knowledge that they could not be excluded from its benefits. The same is true of health and education where the benefits are often for the individuals concerned, but there are also the external benefits to firms of having a well-educated workforce and to society in terms of the economic growth that this could encourage and, the more enlightened individual members of society are, the greater is the possibility of that society being a harmonious one. In the case of health, free immunisation programmes protect the individual from a communicable disease, but also prevent that disease from spreading to others. In all of these cases the free-rider problem exists, but the market is unable to solve the problem. Therefore, government intervention is required to compel individuals to contribute to their provision in the interests of society as a whole.

Neo-classical economists accept that the correction of these market failures is required, but they adopt a minimalist approach. That is, the government should intervene, but only to the minimum extent by first, attempting to improve the functioning of the market by promoting market institutions and second, by always preferring market-based instruments such as taxes and subsidies, over non-market interventions such as regulation or credit allocation. The overarching role for government is seen to be the provision of an economic environment within which markets can operate and resources can be allocated through price signals. Governments should not enter directly into production as they suggest that this will result in firms that are insulated from the rigours of competition, lead to inefficiency and stifle the market. Leftwich has argued that the purpose of structural adjustment programmes has been 'to shatter the dominant post-war paradigm and overcome the problems of developmental stagnation by promoting open and free competitive market economies supervised by minimal states' (Leftwich 1994: 366).

Conclusion

Given the number of areas that are open to reform, both in the internal economy and the external sector, there inevitably arises a conflict between the instruments and objectives of the reforms from the interaction between them and the often incompatible nature of the policies pursued. For example, low interest rates for domestic investment may well affect the exchange rate such that the terms of trade are pushed against the domestic economy, causing a balance-of-payments problem that could be rectified by pushing up interest rates, causing an internal–external conflict. Thus, merely advocating free markets, free trade and minimal government intervention does not necessarily lead to economic growth or economic development, because the economy is much more complex than that.

In addition, as explained in chapter 5, good governance is demanded under the conditions of a structural adjustment loan and many countries have begun a process of democratisation just to appease the Bretton Woods Institutions so that funding can continue. This was stated by Johnson of the World Bank in 1994 in the following terms: 'When the government's legitimate authority is inadequate, program designers must advise the government to obtain it' (Johnson 1994: 399). He then goes on to argue that if a government does not have the political legitimacy to implement the austerity measures that a structural adjustment programme demands, then the costs to that government for the purpose of coercion of the population will be high. However, what this argument fails to grasp is that any democratically elected government which implemented an austerity package could be ousted from power at the ballot box by an opposition promising populist measures. Leftwich has suggested that a structural adjustment loan in its true form cannot be implemented by any government other than an authoritarian one (Leftwich 1994: 367).

The recipe that we are left with is one that could result in disaster. The same economic policies that were introduced in the advanced capitalist economies of the West, and produced unemployment, poverty and increases in the unequal distribution of income and wealth, are capable of producing much worse outcomes in underdeveloped economies lacking in social safety mechanisms and beginning from a position of weakness in world markets. It is to these effects that we turn now in part II.

Part II

Adjustment in practice

Introduction

> The advocates of adjustment saw cuts in public services, increases in unemploy-
> ment and deepening poverty as painful but unavoidable by-products of economic
> modernization. They hoped for a trade-off: short-term social cost set against
> long-term economic gain. But what they did not foresee was that the social
> impact could itself frustrate the desired economic effect.
>
> (UNRISD 1995: 42)

This part comprises four chapters which outline the major impacts of SAPs. We
must stress at this stage that it is artificial to separate out the impacts in the way
we have. For example, environmental impacts relate to economic well-being
which may well elicit a political response and change the social organisation of
production. However, we have divided the impacts into macro-economic,
social, political and environmental chapters for clarity. In order to demonstrate
the inter-connectedness of these processes we have included a variety of case
studies.

Chapter 3 examines the economic impacts of each major policy measure
described in chapter 2. While the impacts have varied they have generally not
been as successful as hoped by the lenders and, in many cases, have resulted in a
worsening economic position. Chapter 4 goes beneath the surface of macro-
economic statistics to look at the complex ways in which different social groups
are affected by the adjustment process. Like any political decision, adjustment
creates winners and losers and it is important that these individuals and groups
are identified within aggregate trends. In particular, the poor are hit hardest by
structural adjustment and of these it is women who bear the brunt of this hardship. The chapter also reviews the policy responses to these differential impacts
and assesses their likelihood of alleviating poverty.

Chapter 5 examines the political processes surrounding the generation and
implementation of SAPs. Although difficult to do in practice this chapter distin-
guishes between the politics of policy formulation and that of policy
implementation. We argue that both phases tend to be accompanied by
increased authoritarianism and decreased accountability on the part of the

lenders. This leads us into the final section on good governance whereby the latest political agenda is to enable market friendly interventions.

Chapter 6 examines an issue which came late to the attention of scholars and policy-makers, but was quickly realised by the local people affected. This was the environmental impacts of SAPs. The formulation of SAPs took place in an environment where 'sustainability' was superseded by the rapid need to improve the balance-of-payments position. Lenders and policy-makers focused on the need to raise exports and cut government expenditure. The overall result was to enhance natural resource exploitation for export or to increase the hardship of the marginalised who, in turn, increased their utilisation of resources in order to survive. The realisation that SAPs are not ecologically sustainable fostered a range of political responses which link into the discussions of the final part.

3 The macro-economic impacts of structural adjustment lending

Bob Milward

Introduction

The World Bank has defined structural adjustment lending as 'non-project lending to support programmes of policy and institutional change necessary to modify the structure of the economy so that it can maintain both its growth rate and the viability of its balance of payments in the medium term' (Stern 1993). In 1988, the World Bank argued that the structural adjustment lending programme had been initiated as a response to fundamental disequilibrium resulting from domestic policy weakness and the effects of the second oil shock (World Bank 1988). However, it has been argued that attempts by the BWIs to promote economic recovery and economic growth have been flawed by their inappropriate approach and the mix of policies required under structural adjustment and conditionality. As such, external resources required to close balance-of-payments gaps and increase financial inflows have not in general increased as a result of the policy requirements of the World Bank and IMF (cf. Griesgraber and Gunter 1996b). This chapter examines these broad economic impacts of SAPs and tracks through the major policy areas identified in the previous chapter. We do this by examining aggregate trends across regions in the first section while in the second we disaggregate by policy reform area.

General trends in economic performance

This section provides an analysis of the performance of a sample of key adjusting countries in Africa and Latin America. Table 3.1 illustrates those policy areas where conditions have been most often applied in terms of structural adjustment loans and Greenaway and Morrissey (1993) point out that exchange rate policy has a relatively low subjection to conditionality here because this is the policy variable most associated with the IMF, and as many countries in receipt of structural adjustment lending are also recipients of stabilisation loans from the IMF, the exchange rate is targeted there.

From tables 3.2 and 3.3 it is evident that the performance of these economies has varied widely during the period of structural adjustment and conditionality. The Latin American economies appear to have fared much better

Table 3.1 The conditionality content of structural adjustment loans (percentage of total loans with conditions in various policy areas)

Policy	Sub-Saharan Africa	Highly indebted	Other
Exchange rate	30.8	18.2	0.0
Trade policies	76.9	90.9	62.5
Fiscal policy	61.5	72.2	56.3
Budget/public spending	69.2	50.0	37.5
Public enterprises	61.5	54.5	43.8
Financial sector	38.5	36.4	43.8
Industrial policy	53.8	9.1	25.0
Energy policy	7.7	13.6	50.0
Agricultural policy	76.9	40.9	37.5
Other	23.1	9.1	12.5

Source: Greenaway and Morrissey 1993: 244

in terms of the internal indicators with large increases in per capita income, particularly in Argentina and Chile, and reductions in infant mortality. However, these economies continue to suffer from high levels of inflation and with current account deficits, with the exception of Chile. They also remain highly indebted despite the reforms that have been undertaken. The African economies here have not progressed over the period and indeed it would appear that Burundi and Kenya have actually regressed, with Ghana standing still. Senegal has increased its per capita GDP, but not to any great extent. It would appear that those economies of Latin America that required external adjustment have not had these problems addressed and continue, in the main, to have balance-of-payments problems with highly inflationary economies and the African economies have neither internal nor external equilibrium, but both Latin America and Africa are highly indebted. This would suggest that for the most part the neo-classical solutions have not worked and indeed in many cases have made the situation worse.

Policy reforms and outcomes

This section disaggregates the impact of SAPs by those policy areas outlined in chapter 2. We do this through selected case studies since this permits attention to the unique circumstances operating in each country and, therefore, the ways in which relatively standardised adjustment packages unravelled in different ways.

Table 3.2 Total fund credit and loans outstanding (millions of SDRs)

	Africa	Burundi	Ghana	Kenya	Senegal	Argentina	Brazil	Chile	Peru
1986	7,185.06	18.08	642.26	375.90	236.25	2,240.85	3,679.86	1,088.25	595.51
1987	6,326.39	14.48	610.94	282.89	241.64	2,716.24	2,802.88	1,032.38	595.44
1988	6,118.59	24.12	566.36	338.44	236.54	2,733.00	2,476.82	982.58	595.42
1989	6,117.95	30.54	561.16	316.12	240.42	2,358.82	1,843.36	966.46	576.98
1990	5,762.75	29.96	523.39	338.88	220.95	2,167.20	1,279.68	812.90	530.46
1991	5,896.24	34.16	583.12	344.84	228.90	1,735.92	865.14	669.38	493.41
1992	5,727.21	47.39	537.83	286.08	197.36	1,682.80	581.40	525.02	458.72
1993	5,871.46	44.40	537.30	264.34	177.81	2,562.44	221.02	346.50	642.69
1994	6,531.56	40.13	479.70	277.25	205.43	2,884.69	127.50	199.48	642.69
1995	7,065.30	34.16	436.24	251.45	233.31	4,124.39	95.23	/	642.69
1996	7,457.32	28.18	377.35	234.51	226.46	4,376.04	47.02	/	642.69

Source: IMF 1997: 39

Note: / represents data not available

Table 3.3 Economic indicators

	Burundi	Ghana	Kenya	Senegal	Argentina	Brazil	Chile	Peru
Ranking (1988)	13	30	23	33	84	74	69	63
Ranking (1997)	4	31	2	41	105	96	101	80
GNP per capita ($)								
(1986)	240	390	300	420	2,350	1,810	1,320	1,090
(1995)	160	390	280	600	8,030	3,640	4,160	2,310
(1997)	180	370	330	550	8,570	4,720	5,020	2,460
Av. Ann. Growth (%)								
GDP (1965–86)	1.8	-1.7	1.9	-0.6	0.2	4.3	-0.2	0.1
GDP (1985–95)	-1.3	1.4	0.1	–	1.8	-0.8	6.1	-1.6
GDP (1990–97)	-3.7	4.3	2.0	2.4	4.5	3.1	7.2	6.0
Life Expectancy								
(1986)	48	54	57	47	70	65	71	60
(1995)	49	59	58	50	73	67	72	66
Ann. Av. Inflation (%)								
(1980–86)	6.4	50.8	9.9	9.5	326.2	157.1	20.2	100.1
(1985–95)	6.1	28.6	13.0	3.7	255.6	875.3	17.9	398.5
Current Account Balance (% GDP)								
(1980)	–	0.7	-12.1	-12.8	-6.3	-5.5	-7.1	-0.5
(1995)	0.6	-6.5	-4.4	0.1	-1.4	-2.6	0.2	-7.4
Infant Mortality (per 1000 – under 5)								
(1980)	121	100	72	91	35	70	32	81
(1995)	98	73	58	62	22	44	12	47
(1996)	176	110	90	88	25	42	78	71

Continued:

	Burundi	Ghana	Kenya	Senegal	Argentina	Brazil	Chile	Peru
General Government Consumption (% GDP)								
(1980)	9	11	20	22	–	9	12	11
(1995)	12	12	15	11	–	17	9	6
(1997)	10	10	17	10	–	16	9	11
Current Account Balance ($m)								
(1970)	2	-68	-49	-16	-163	-837	-91	202
(1980)	–	30	-878	-387	-4,774	-12,831	-1,971	-101
(1986)	-38	-43	-42	-284	-2,864	-4,930	-1,091	-1,055
(1995)	-6	-414	-400	3	-2,390	-18,136	157	-4,223
(1996)	-6	-324	-74	-58	-4,136	-18,136	-2,921	-3,607

Source: World Bank 1988 and World Bank 1997

Fiscal policy and management

The use of fiscal policy in the neo-classical paradigm suggests that a system of taxation and subsidy is inefficient, and as such the approach preferred in a structural adjustment package is that of lowering taxes to increase individual incentives and of eliminating subsidies to raise efficiency through greater competition. In addition, it is argued that state investment programmes, funded out of taxation, are a highly inefficient use of scarce resources.

For example, Burundi faced unstable coffee prices which had a direct effect as coffee is of major importance to government revenue. Given a high level of state spending on higher education and hospitals and a high degree of military expenditure, reductions in revenue from coffee had a large impact on the state budget. In addition, the government in Burundi was giving direct and indirect subsidies to the public enterprises, which compounded the problems of inefficiency in the investments in public projects.

The reform package consisted of three main public expenditure policies. First, the government agreed to reduce the size of the public investment programme and to produce an investment plan that was compatible with the availability of resources. Second, budgetary reform involving the introduction of a unified budget system which would identify and separate current revenue, current expenditure and capital expenditure and the preparation of a three-year rolling programme for public expenditure. Third, the strengthening of the agencies responsible for preparation of the annual budget at the ministry of finance and to establish a system for the monitoring of project implementation.

In the early 1990s, under phase III of the reform programme, the government was attempting to increase the funding of investment-related recurrent costs and to promote quality-based services, thus improving the composition of public expenditure. The public investment programme has scaled down the size of projects and promoted more realistic sectoral objectives that take account of resource and debt service constraints. Approximately 75 per cent of capital expenditure in the public sector expenditure programmes goes to the economic infrastructure, of which 30 per cent goes to roads, 30–40 per cent to the rural sector, whereas social infrastructure receives 10–13 per cent, primarily for health and education (Englebert and Hoffman 1996). However, this represents a substantial problem for several sectors: agriculture accounts for 60 per cent of the underfinancing with large items being neglected, including the maintenance of livestock, infrastructure, forestry and irrigation facilities. Shortfalls in the health sector represent approximately 14 per cent of total underfinancing and include medicines, equipment and vehicles. The underfinancing of road maintenance represents 16 per cent and the remaining 10 per cent is spread across other sectors (Englebert and Hoffman 1996). The previously relatively large public investment projects in the productive sectors of the economy have gradually shifted towards smaller investments that are targeted primarily at improving the quality of the social services. Government expenditures were insufficiently restructured however, which has tended to reveal a weak commitment to the reduction of public expenditure. The wage bill has been excessive in relation to expenditures on goods and services and the accumulation of large counterpart funds by the public sector has tended to undermine the principal goal of the programme, to reduce the role of the state in the economy.

It has been argued that fiscal policy in Kenya has been the single most critical variable in macro-economic management, and yet, at the heart of its fiscal problem is the inability to control expenditures and not its inability to generate sufficient revenue. The tax to GDP ratio hovered around 22–23 per cent for most of the 1980s, whilst the ratio of recurrent expenditures increased significantly. The inability to control expenditure was due in part to the structure of expenditure and in part to a general lack of discipline in expenditure allocation and execution. There has been a downward inflexibility characterised by the civil service wage bill that was 6 per cent of GDP during the 1980s. There was also upward pressure from the growing demands of teachers' salaries and interest payments, particularly on the domestic debt of the government and the parastatals. The government was also thwarted by the absence of clear expenditure priorities and the political will to control expenditure. One result was that expenditure ceilings set by the treasury were repeatedly violated. There was also a proliferation of government functions with a sharp increase in the number of departments and divisions, making co-ordination cumbersome. In 1985 the government initiated a budget rationalisation programme designed to introduce better planning and discipline into resource programming and expenditure, seeking greater balance between developmental and current spending, and between wage and non-wage spending. The budget rationalisation programme also sought to target scarce managerial and financial resources on a smaller number of critical 'core' investment projects. However, this did not improve resource allocation, and between 1981 and 1986, non-wage operations and maintenance expenditures fell from 36 per cent to 26 per cent of total recurrent expenditure and, after a further budget rationalisation programme, they declined further to 22 per cent (Swamy 1996). The government increasingly relied on the development budget, particularly the donor-financed component, to finance recurrent expenditures. Rather than raise additional revenue relative to GDP, the tax reform programme sought to primarily simplify the tax structure. It lowered the high tax rates that may encourage evasion and discourage investment, strengthened tax administration and increased the use of user charges, particularly in health and education. It focused on import tax and company taxation, with the government reducing company income tax from 52 per cent in 1982 to 37.5 per cent in 1992 and reducing the average unweighted import tariff (Swamy 1996). Tax coverage was also broadened to cover more products and services. In 1990, the government sought to improve tax collection by introducing the tax modernisation programme, which included attempts to computerise the tax system.

In Latin America in the 1980s, most governments implemented policies designed to eliminate their fiscal deficits. By the early 1990s, these large fiscal deficits had disappeared in the major economies and some were running surpluses.

In 1976, Argentina entered into a five-year agreement with the IMF which entailed a full-scale neo-classical policy conditionality. The programme consisted of the liberalisation of external trade and capital markets, the devaluation of the nominal exchange rate, the elimination of domestic subsidies, the privatisation of public enterprises and the raising of public sector prices and domestic interest rates. This attempt to stabilise and to adjust the economy in the policy frame-work of freeing up markets and allowing individuals to initiate economic decisions appeared, initially, to be successful in terms of the main economic indicators. Real output grew by more than 12 per cent between 1977 and 1981, the rate of inflation fell from 443.2 per cent in 1976 to below 150 per cent on average for 1977–81 and a current account surplus was generated until the final quarter of 1979. However, Argentina entered into a steep recession in 1981–82, as external debt increased from $6 billion in 1976 to $14.4 billion in 1981, accompanied by capital flight. In addition, the current account deficit reached 3.8 per cent of GDP in 1981. The outcome was a reversal of the real output growth generated in the previous four years. A phase of gradual adjustment began in 1983 as the government signed a stand-by agreement with the IMF under similar conditions to the programme of 1976. The outcome of this was a decline in real wages of 32.5 per cent, an increase in the rate of inflation to over 600 per cent and a reduction in output of 4.4 per cent, all by 1985 (Ruccio 1991).

For Argentina, the decade of the 1980s was a period of stagnation and financial crisis in the public sector. The fiscal deficit/M3 ratio[1] was 26.2 per cent prior to the crisis and in the years that followed it often exceeded 100 per cent. With a low demand for government bonds and the rationing of foreign credit, the government was forced to monetise a large part of the deficit with a huge increase in inflation tax. However, this led to a fall in the demand for money, contributing to a worsening financial position. Attempts were made to raise government revenues through increases in the overall tax burden, but all failed. Thus, fiscal adjustment fell on public expenditures and in particular on public investment. The public investment GDP ratio fell from 6.5 per cent in 1980 to 3.1 per cent in 1990 (Chisari *et al.* 1996). The knock-on effect of this was a decline in private sector expenditure on capital. Structural reforms since 1990 have included a law regulating the activities of the central bank which prohibited the monetary financing of the fiscal deficit and a large privatisation programme which aided considerably the reduction in the deficit. In addition, the tax burden increased by approximately 5 per cent between 1990 and 1995. This was achieved through a broadening of the tax base and improvements in administration. The fiscal deficit/M3 ratio fell from 136.8 per cent in 1990 to 2.7 per cent in 1992 (Chisari *et al.* 1996). However, public investment has remained very low indeed and the privatisation programme has had a major role in closing the fiscal gap.

Monetary policy

Neo-classical theory suggests that monetary policy should be used to address the problem of inflation through control of the monetary base and involves reform of the control that is exercised by the central bank over all financial intermediaries. The idea is basically that of the so-called 'monetarists', whereby reductions in the rate of growth of the money supply result in reductions in the rate of growth of the price level. This theory was adopted by many developed economies in the late 1970s and early 1980s, but was quickly abandoned as a major economic strategy due to the effects of reductions in aggregate demand and the subsequent appearance of mass unemployment. However, it has remained as an important policy instrument of structural adjustment programmes.

In Ghana, monetary policy appeared to become more manageable with improved fiscal discipline. As the economic recovery programme was implemented, the monetary base began to stabilise and the rate of inflation fell slightly. However, the initial success of the stabilisation programme complicated monetary management in new ways. External financial support rose dramatically, allowing a replenishment of depleted foreign exchange reserves. Although the build-up of reserves was partly a policy choice, it was also necessitated by the government's financing of local expenditures. For example, in 1988 and 1989, domestic spending in excess of domestic revenue was financed by the conversion of foreign aid. Thus, foreign assets held by the Bank of Ghana became an important source of monetary growth, replacing the fiscal deficits of a few years earlier. However, since 1989, domestic credit policy has been used aggressively to offset the growth of foreign assets. Initially the policy generated a small surplus which was used to repay the government's debt to the Bank of Ghana to offset monetary growth and in subsequent years repayments were increased and domestic credit extended by the Bank of Ghana was reduced. One outcome of this was that inflation fell to single figures by the end of 1991.

Monetary reform in Senegal was threefold: first, managing overall demand to correct external imbalances; second, gaining a competitive advantage by keeping the level of inflation below that of its major trading partners; and third, strengthening the management of liquidity and the supervision of the banks. The government addressed the first two of these objectives primarily by pursuing restrictive credit policies and it implemented a more comprehensive set of measures in the banking sector, including the introduction of market-determined interest rates. In September 1989 Senegal, and other members of the West African Monetary Union, adopted a comprehensive reform of monetary policy instruments that replaced the administrative controls over money and credit with an indirect and market-oriented system of monetary instruments. In particular, the preferential rediscount rate was abolished and the central banks' rediscount rate was set above money market rates, whilst banks were given more flexibility in determining their rates on deposits and loans. In addition, rigorous controls were placed on state guarantees for borrowing by public and private enterprises with the system of sectoral credit being eliminated. Restrictive monetary policy has helped Senegal's external account but has severely constrained domestic credit. From 1986 to 1991 credit to the non-government sector increased by an average of only 0.7 per cent per annum and in real terms it declined substantially. The money supply followed a similar trend, although the decline was less severe. One result of monetary reform has been the liberalisation of the interest rate structure. By eliminating the preferential discount rate, which formerly applied to the agricultural sector, the export sector and residential construction, the government has kept lending rates for prime borrowers at 16–18 per cent while maintaining official inflation at less than 3 per cent. Thus, the measures have kept inflation low, curtailed overall domestic demand and restricted government intervention in the allocation of credit. Yet this has not created a foundation for sustainable growth.

The average rate of inflation in Latin America increased 26 times between 1981 and 1990, with those most affected being Argentina and Peru, who suffered hyper-inflation, but also Uruguay and Mexico with high levels of inflation. By 1993 the rate of inflation had been reduced, but was still relatively high in Uruguay and Peru (Mesa-Lago 1997). In Brazil, the rate of inflation fell from 228 per cent in 1985 to 58 per cent in 1986, but was at nearly 1,000 per cent in 1988 (Green 1995).

The Economic Solidarity Pact in Mexico, introduced in December 1987, was designed to quickly reduce the rate of inflation and relied on multiple economic instruments, including a fixed exchange rate and an incomes policy. In addition, domestic interest rates were set above international rates and a restrictive monetary policy was adopted. The money supply contracted between 1988 and 1989 as real interest rates rose sharply. However, the rate of inflation continued to rise faster than nominal wages (Nazmi 1997).

Trade policy

As we observed in chapter 2, neo-classical trade policy involves the elimination of restrictions on trade, including tariff and non-tariff barriers, to achieve the optimum level of production and consumption. In addition, an outward, rather than inward, orientation of production is required under structural adjustment to achieve greater competition to enable increases in the efficiency of local output.

A study of the protection system in Burundi prior to adjustment identified a wide spread in customs tariffs (68 to 336 per cent), quantitative restrictions (quotas for most products and a ban on imports that competed with locally manufactured products), import regulations, foreign exchange controls and compulsory advance deposits with the central bank. Imports were also subjected to three different kinds of tax: customs duties, a statistics tax and a transactions tax. In addition, all commercial imports required licences. The reforms were sequenced, seeking to liberalise the trade regime. Import licences were to be granted automatically for most products, with the exception of cotton textiles, glass and pharmaceutical products. Regulations on importers were to be eased to facilitate competition. A simplified tariff structure was also proposed to reduce the number of duties from three to two, the number of duty rates from fifty-seven to five and the spread from 50 per cent to 15 per cent in 1986 to 40 per cent to 20 per cent in 1989 (Englebert and Hoffman 1996). The second phase of the programme called for a reforming of the tariff structure further, liberalising the import of locally manufactured products, increasing the ceiling on import licences and requiring that the central bank pay interest on the FBu 10 million that foreign importers were obliged to deposit as a guarantee against illegal business practices. To promote exports, the programme called for the adoption of a simplified drawback procedure for import taxes and the simplification of administrative procedures for businessmen travelling abroad. Tariff reductions and the liberalisation of import licensing were implemented much slower than anticipated and while quotas were abolished and the tariff structure was simplified, tariffs on luxury goods remained and those on intermediate and capital goods fell more slowly than had been envisaged in the programme. At the same time, the transactions tax rate was increased as part of a revenue enhancement package of the stabilisation progamme and cancelled out some of the impact of lower tariffs. Overall, the adjustment programme reduced effective protection mainly by eliminating non-tariff barriers, but not by as much as had been originally envisaged. Until the mid-1990s, importers continued to complain about administrative harassment and constraints as all commercial imports required licences. Hence, the impact of trade reform has been limited.

Import substitution behind protective barriers through state intervention were the main principles of industrialisation in Kenya. In 1979 it was recognised that the industrialisation strategy would have to move to a more outward-oriented approach. Thus, between 1980 and 1984 it was envisaged that there would be phased replacement of qualitative restrictions with equivalent tariffs and a subsequent tariff rationalisation to provide a more uniform and moderate structure of tariff protection. But the Kenyan government halted the reform and introduced import controls for some items when the balance-of-payments position deteriorated, and during the exchange rate crisis of 1982–84 tariffs increased by 10 per cent across the board. In 1988 yet another attempt was made to liberalise imports through tariff reform as the highest rate was reduced from 135 per cent to 60 per cent and the number of categories fell from twenty-five to twelve, with tariff rates on non-competing imports being lowered (Swamy 1996). During much of the 1980s, Kenya had a significant anti-export bias on non-traditional goods. In 1988, the government introduced the 'manufacturing-under-bond scheme', allowing customs authorities to waive import duties and taxes on imported materials used as inputs for the production of export goods, and in 1990 a more general import duty/value added tax exemption scheme was introduced. This was complemented by a regulatory reform designed to centralise and consolidate several licensing procedures for exports and general trade with simplified procedures for new investments. 1990 saw the construction of an export-processing zone near Nairobi with exemption from tax for ten years, unrestricted foreign ownership and employment of foreigners and complete control over foreign exchange earnings.

Whilst Latin America's exports doubled from $78 billion in 1986 to $153 billion in 1994, there was an enormous increase in imports following trade liberalisation and generally overvalued exchange rates. The major problem appeared in Mexico which ran a trade deficit of $24 billion in 1994 (Green 1995). Free trade has also led to a proliferation of trade agreements of both a regional and global nature in terms of the General Agreement on Tariffs and Trade (GATT), the North American Free Trade Agreement (NAFTA) and of Murcosur, which involves Argentina, Brazil, Uruguay and Paraguay.

Trade reform in Mexico began in 1985 with a reduction in the number of tariff rates to five, and the introduction of a maximum tariff that was set at 20 per cent. The government also reduced direct export subsidies, and by 1991 export incentives in Mexico included a tariff exemption on temporary imports and a programme exempting exports from import licences on inputs (Ros *et al.* 1996). In the same year only goods and commodities subject to price controls or international agreements required export permits and export tariffs were set at a maximum of 5.5 per cent. Mexico had also become a member of GATT in 1986 and NAFTA was signed in January 1994. As a member of GATT Mexico agreed to bind its tariff schedule to a maximum of 50 per cent *ad valorem* and also to implement reductions on the tariffs of the majority of its imports to 20–50 per cent (ibid.). However, by 1992 Mexico was generating a trade deficit larger than that of the early 1980s. This has been due in part to increases in imports due to the import liberalisation, and this has been particularly true of consumer goods as nearly 80 per cent of the decline in the non-oil trade balance is associated with the increase in consumer exports after 1987 (ibid.).

Similarly in Argentina, a trade account surplus of $8.3 billion in 1990 turned to a deficit of $2.6 billion in 1992 and of $3.7 billion in 1993 as imports increased from $4.1 billion in 1990 to $14.9 billion in 1992 and to approximately $16.4 billion in 1993 (Chisari *et al.* 1996). Again this was the outcome of the liberalisation of trade that included the elimination of trade restrictions and the reduction in tariffs.

Prices and market deregulation

The deregulation of prices and markets is essentially the crux of the neo-classical paradigm and hence is at the heart of all structural adjustment programmes. It entails the removal of government intervention in all aspects of price determination and, therefore, the forces of supply and demand become the determinants of prices in all markets. This follows the neo-classical theory in which prices that are determined by forces other than those of supply and demand will lead to inefficiency in the allocation of scarce resources.

Prior to the adjustment programme in Burundi, all prices of imported and locally manufactured goods were subject to controls exercised by the Ministry of Commerce and Industry. Prices were set on a cost-plus basis with the manufacturer receiving a net profit margin of 10–20 per cent. The government decided to deregulate prices at the outset of the programme and to address both producer and consumer prices. They agreed to let foodcrop producer prices to be market determined and established prices for cash crops at levels sufficient to provide incentives for increased production and quality, taking into account the evolution of international prices. In addition, the government chose to liberalise the consumer price control system. The automatic system for adjusting producer prices was not implemented under the structural adjustment credit I, and although measures to improve the quality of certain export crops and to increase their production were partially successful, progress towards improving international marketing arrangements and maintaining competitive producer prices has not been satisfactory. Measures to liberalise the rice sector and to restructure the sugar complex were implemented as planned. Coffee prices remained below the 1980 level in real terms. Nominal producer prices rose, but decisions about production and producer prices continued to be made administratively and compulsory controls on plantations prevented farmers from shifting to crops with higher returns (Englebert and Hoffman 1996).

Despite attempts at deregulation, markets for most crops remained monopolistic or quasi-monopolistic as public enterprises continued to dominate the main sectors. Virtually no policy change occurred at the farm gate level and the few changes that did take place were sterilised partly by the monopolistic nature of these markets. However, the boom in non-traditional exports (rice and tobacco) offers a contrast to the lack of success in more visible areas of tea, cotton and coffee. The liberalisation of rice prices and the introduction of private hullers led to price increases of approximately 20 per cent between 1989 and mid-1990, turning Burundi from a net importer to a net exporter of rice. Increases in the production of tobacco also reflect price incentives, as well as the technical and input support offered by the Burundi Tobacco Company to farmers under contract. In May 1992, based on the lessons from these structural adjustment episodes, the conditionality of a further structural adjustment credit included several new measures to address market and price deregulation. The regulatory framework for smallholders was simplified through, among other things, the elimination of government constraints on mandatory crops and the quality and types of inputs to be applied. Also, government-administered producer prices for traditional export crops and agricultural inputs were eliminated. Public enterprise monopoly rights for purchasing, marketing and processing agricultural inputs were abolished. Finally, a moratorium was imposed on all new public investments in tea, rice and palm oil sectors and transparent criteria for allocating available public lands and resolving land ownership disputes were established.

The World Bank prepared two sectoral reports for Senegal in 1981 and 1989 to assist the government in formulating an overall framework for reform. In 1986, public enterprise reform became part of the government's medium-term economic and financial rehabilitation programme, supported by three structural adjustment loans, a financial sector adjustment loan and a technical assistance project. The objectives were to restrict government intervention in enterprises that could be more effectively managed by the private sector and to improve the government's ability to manage those public enterprises that remained under state control. In 1985, the government strategy called for the privatising and liquidating of selected public enterprises, improvement of management and the performance of those that remained in the public sector, simplifying sectoral control procedures and improving portfolio management information. Since 1990, the strategy has shifted from attempts to privatise non-performing public enterprises, to efforts to start with profitable enterprises and strengthen the process of privatisation.

Public enterprise reform in Senegal has been plagued by significant delays. Over the entire reform period the government has liquidated twenty-one public enterprises and totally privatised twenty-six others, together representing 42 per cent of the total number of public enterprises in the sector. However, in terms of assets and government equity, they represent only 19 per cent and 11 per cent of the sector respectively. The disproportionate percentages reflect the fact that the three utilities of electricity, water and telephones were not part of the divestiture programme, and they alone account for 33 per cent of the sector's assets and 46 per cent of government equity (Rouis 1996).

The labour market

Neo-classical theory in the area of labour markets is again concerned with the elimination of regulations that impinge upon the free operation of the forces of supply and demand. In particular it is concerned with any regulations that decrease the incentives of employers to take on workers and therefore have the effect of reducing the demand for labour. The forces of supply and demand, as illustrated in figure 2.2, operate in the same manner in the labour market as in any other commodity market, but the equilibrium position produces an equilibrium wage. Hence, interference in this market will distort the outcome and produce a non-equilibrium wage. According to the theory this will be a below-optimum outcome and could lead to unemployment of labour as potential employers either cut back production or introduce more capital-intensive methods of production to reduce their wage bills.

The labour market in Ghana is segmented, with an informal sector (rural and urban) and an organised modern sector. In the informal sector, wages are determined by market forces, but in the organised sector the government is the wage leader, accounting for approximately two-thirds of total employment. Since the mid-1980s, the government has attempted to restructure the size of the public sector workforce and its remuneration. Some progress towards retrenchment has been made, but a severe shortage of data makes assessment of progress difficult. Despite pervasive overemployment, public sector wages have increased substantially in real terms during the period of adjustment. Between 1983 and 1988, average real earnings in the public sector more than tripled. Much of the increase sought to compensate for substantial inflationary erosion that had occurred prior to adjustment. Between 1988 and 1992, the government attempted to link wage increases to a measure of productivity. Average real earnings were held more or less constant, while salary differentials were permitted to widen across grade levels and occupational categories. In 1992, organised wage demands intensified in the months before the elections and in October of that year the government agreed to increase average government service pay by approximately 80 per cent. Average real wages exploded by 500 per cent since adjustment or 20 times the gain in per capita GDP (Leechor 1996). Wage rates and the quality of labour represent a major part of business decisions, particularly those of foreign companies. Wage rates in Ghana remain competitive by international standards, but this wage advantage is offset by relatively low levels of education and skills of the average worker.

Before the reforms of 1986, the modern labour market in Senegal had been heavily regulated by the government and suffered from a highly antagonistic system of industrial relations. The government regulated all labour practices and wage rates through labour law, which included the labour code, the 1981 collective bargaining system and the civil service statutes. These regulations led to higher production costs, limited investment, low productivity and low levels of job creation. The effect was that the government's income policy was at odds with its internal adjustment strategy. Modern sector wages in Senegal are largely influenced by the wages that are paid by the state and in 1986 the government employed 68,000 of the modern sector, which employed 131,000 in total. Real wages in the civil service and in the modern sector declined between 1980 and 1985, but remained high relative to other countries of the region. However, to compensate for this decline, the government progressively raised benefits and wage supplements, which by 1986 made up 25–30 per cent of the wage bill. Svejnar and Terrell found that for the industrial sector as a whole, total factor productivity declined annually by 5–7 per cent between 1980 and 1985 (Svejnar and Terrell 1988). Employers relied heavily on temporary, rather than permanent, employees due to the regulations concerning the laying-off of workers which required government approval and was time-consuming. In 1986 the government established a national employment fund and created an agency to ease the transition of redundant workers and to encourage voluntary departures from the civil service. A World Bank appraisal of this programme concluded that the cost effectiveness of the employment fund had been diverse. Between 1988 and 1991 only 1,500 jobs had been created at a cost of $11,000 each.

The financial sector

Although Burundi's financial sector is reasonably diversified, the government participation in the equity of the financial institutions is quite pervasive. The government holds a 42 per cent share in the capital of the two largest commercial banks. However, financial intermediation remains low, due to the low per capita income of the country, the low monetisation of the economy and a lack of competition among the financial institutions. During the period of the reforms the government attempted to improve its credit and monetary policy in the financial sector to make financial intermediation more efficient and to improve resource mobilisation. The first phase of the adjustment programme proposed raising the authorised credit ceiling of the commercial banks by more than 300 per cent and promoting small and medium-sized enterprises through a reinforcing of the guarantee fund. In addition, an *ad hoc* committee was formed with the remit to address the discrimination that was implicit in the distinction between discountable and non-discountable credit and the structure of interest rates. In the second phase there was an attempt to liberalise interest rates and to supplant credit rationing with a more indirect way in which to manage credit and liquidity by establishing a simplified interest rate structure with only two discount rates. Implementation was slow and incomplete as overregulation and oligopolistic behaviour limited competition among the financial institutions. The government and the central bank introduced a more flexible interest rate policy through the linking of interest rates to the rates obtained during the treasury certificate auctions. However, interest rates did not become fully liberalised.

The financial sector in Kenya was fairly well developed in the mid-1980s and the government amended the banking law in 1989 to narrow the regulatory gap between classes of financial institutions by increasing the minimum capital requirements and by attempting to strengthen the central bank's technical and managerial capacity to examine, monitor and supervise the financial system as a whole. However, the reform measures had a minimal impact as it was estimated that eleven banks and twenty near-bank financial intermediaries were in distress in 1992 and these accounted for approximately 60 per cent of total assets. In addition, interest rate liberalisation failed to reduce the rigidities that existed in the allocation of credit.

Commercial bank deposits in sub-Saharan Africa average 16 per cent of GDP, but in Ghana this is only 8 per cent and the failure to mobilise resources has restricted the availability of credit and raised the cost of investment funds which has had pervasive repercussions for the economy as a whole. Since the adjustment process began in 1983, there has been a decline in the level of available credit, due in part to the direct control of domestic credit up to 1992 and, as a large part of credit is allocated to the publicly owned enterprises, the private sector has experienced a credit deficiency.

Probably the most successful of the reforms undertaken in Senegal has been the reform of the banking sector. In October 1987 the government consulted with the IMF, the World Bank and its bilateral donors with the aim of restructuring its ailing banking sector. The outcome of these discussions was a set of measures that included the closure of the distressed banks, reform of credit policies, the recovery of bad debts and a reduction in abusive practices. In June 1989 the government adopted a capacious strategy of restructuring which called for the maintaining of only those banks that could become profitable and of privatising banks such that government interference would be reduced.

Privatisation

The Kenyan government held equity in 250 commercially oriented enterprises producing goods and services for profit. The single largest economic activity of the parastatals was in manufacturing and the parastatals accounted for more than a third of the government's net lending and equity operations. Several attempts were made to restructure the state sector and in late 1991 the government announced its intention to divest its interest in 207 enterprises whilst retaining ownership of thirty-three strategic enterprises. However, by 1995 only five of these had been privatised or brought to the point of sale.

Ghana has more public enterprises than all other African countries with the exception of Tanzania and they have a presence in virtually all sectors of the economy. They have substantial commercial advantages over private sector enterprises and the government has continued to be ambivalent about public sector reform. Although the government appeared to have little intention of changing its role in commercial activities in the first few years of adjustment, it did acknowledge the existence of an excessive amount of labour and the poor performance of the public sector *per se*. Therefore, rather than allowing competition from private companies, the government attempted to improve the efficiency of the public enterprises. However, as the losses of the public enterprises rose and their burden on the budget increased at the end of the 1980s, the government began to reassess its position. Attempts were made to sell or liquidate minor enterprises and limited participation of private enterprises was allowed. But even by 1992, very few actual measures had been implemented.

Between 1989 and 1992 Argentina and Mexico sold off 173 state companies for $32.5 billion in both cash and debt relief. Generally throughout Latin America, privatisation has been seen as a means of reducing government deficits and of raising funds for current expenditures. However, whilst Mexico may be in the position of demanding competing bids from the international business community for some of its prized assets, the same is not true of other countries in the region such as Peru. Here the government must accept the first serious offer (Orme Jr 1993).

In Mexico, privatisation raised a total of $13.7 billion in the year 1990–91 which represented just less than a tenth of the government's revenue. Of these the largest sales were of the state telecommunications company (Telemex), which raised $4 billion, and the twelve state banks, which raised $10 billion (Green 1995). Other sales included the two leading airlines, Mexicana and Aeromexico, and the copper mining company CANANEA (Ros *et al.* 1996).

In Argentina, the government sought to privatise a large number of state companies between 1990 and 1993, and in the first two years privatisation revenue represented approximately a fifth of government spending (Green 1995). It has been argued that the most important source of financing for the public sector in Argentina has been the funds from the proceeds of the privatisation programme. Indeed, from the $9.8 billion received by the government, $6.2 billion came from the divestiture of public enterprises and further inflows were generated through the debt-equity swap schemes (Chisari *et al.* 1996). Argentina has been the only economy of the region to privatise its state oil company (YPF) whilst others have preferred joint ventures with transnational companies. According to Green, however, problems have arisen in many of the former state enterprises post-privatisation. Call charges in the telecom sector rose by 60 per cent after privatisation and the sell-off of the state airline (Aerolineas Argentinas) to Iberia of Spain led to strikes and demonstrations when the new owners attempted to sack a third of the employees and as a consequence ran up losses of $550 million in the first three years post-privatisation.

Conclusion

As is so often the case with evaluating the impacts of SAPs it is impossible to generalise. A recent assessment concluded that 'Neither the IMF nor the World Bank have been able to demonstrate a convincing connection, in either direction, between SAPs and economic growth' (Killick 1999: 2). The preceding analysis would bear this out, but what it is unable to do is go beyond the aggregate level and look at the distributional effects of SAPs or to assess their unpriced impacts. It is to these matters that we now turn.

Note

1   M3 is a measure of money supply so that the deficit/M3 ratio refers to the size of deficit to amount of money actually available.

4 Social consequences of structural adjustment

Alfred B. Zack-Williams

Introduction

One of the most ardent criticisms levelled against SAPs is that relating to the neglect of the social sector (Husain and Faruquee 1995: 9).[1] Indeed, there is an *a priori* assumption that both the urban and rural sectors would gain from the economic and financial liberalisation (Sahn *et al.* 1994), that is central to SAPs, as 'the real exchange rate depreciation improves production incentives for tradable goods; agricultural output rises, leading to increased labour demand and higher real incomes of the poor' (Sahn *et al.* 1994: 34). Indeed, it is argued that SAPs would help to improve the terms of trade in favour of the rural inhabitants, the locus of Africa's poor. Sahn *et al.* have argued that the negative impact on the poor of exchange rate depreciation is far less than critics of adjustment would have us believe, since the poor have restricted access to foreign exchange markets; remain heavily immersed in subsistence production and have limited access to subsidies, public employment and services. Sahn *et al.* (1994) focused upon the inefficient and wasteful use of resources, even in the face of severe total budget constraints, the concentration on curative rather than primary health care, as well as targeting of resources on secondary and university education. In their view, these expenditures tend to favour the urban elites, who are the ones who stand to lose in any process of adjustment.

However, in trying to locate the social consequences of adjustments in most Third World countries, particularly African countries, there are a number of difficulties that are immediately encountered by the researcher. Paucity of data is one such factor. For early adjusters such as Jamaica (1977–80), Ghana (1983), Mali (1981) and Malawi (1981), the question of the social impact of adjustments was not a serious item on the adjustment agenda, at least, not before the publication of UNICEF's *Adjustment with a Human Face* (Cornia *et al.* 1987). In the case of Ghana, this led to the setting up of the PAMSCAD (Programme of Action to Mitigate the Social Costs of Adjustment), which we shall return to presently.

The second problem is that of disaggregating the social consequences of SAPs *per se*, from those which are the products of a general crisis of the post-colonial mode of accumulation.[2] Indeed, these 'before-vs-after' approaches, as

Paul Mosley has labelled this argument, cannot by their very nature resolve the issue. In order to solve this methodological problem, some studies have concentrated their analysis 'by evaluating only policies actually changed, and by examining their impact exclusively of other shocks, positive or negative, in the economies' (Sahn *et al.* 1994). Others utilise 'the comparative static' approach, by comparing the performance of adjustment lending countries with that of non-adjustment lending countries (Grosh 1994). However, this approach raises more problems than it solves as Mosley has warned:

> Although overall the performance of the adjustment lending countries was better on all indicators selected (except investment) than that of 'non-adjusting' countries within the low-income and Sub-Saharan groups the growth and export performance of 'non-adjustment lending' countries was superior to that of the countries which received loans.
>
> (Mosley 1994: 71)

Finally, the amount of slippage in terms of the conditionality means that it is near impossible to compare like with like. This slippage is often premised on the fear of the political consequences of the deflationary effects of SAPs on powerful interest groups, such as the urban elite and organised labour (Westebbe 1994; Zack-Williams 1997). In other cases, this reflects the inability of some states to keep up with arrears owed to the IFIs, in which case the latter unilaterally abrogate the agreement.

Given these methodological pitfalls, perhaps the most useful approach to look at the social impacts of SAPs is to try to disaggregate the impact of policies such as devaluation, downsizing of the bureaucracy, cost sharing/recovery, and the lifting of subsidies, on various groups within society. These policies would have differential impacts on various sectors of society; there may be winners and losers. For example, studies by Elson (1989), Laurie (1997), Emeagwali (1995), Sparr (1994), Gibbon and Bangura (1992) and Osaghae (1995) have all pointed to the gender, class and ethnic dimension of SAPs. The first victims of adjustment policies are usually daily-wage workers, who are structured out of employment and rendered as sacrificial lambs at the altar of SAPs.

The effects of SAPs on poverty and well-being

Jean-Paul Azam has identified two major avenues where adjustment policies affect social groups: the distribution of real income by the market; and the provision of public goods by the state (Azam 1994). With regards to the former, we note that one important item on the adjustment agenda is to effect change in the terms of trade in favour of rural producers in Africa. This policy is premised on two major assumptions: that the poor in Africa are located in the rural areas, and by tilting the rural–urban terms of trade in their favour, this would help alleviate poverty. By boosting the level of production in the tradable sector, while concurrently reducing output in the non-tradable, it is assumed that 'firms' in the tradable sector will be induced to take on more workers (Azam 1994).

Nonetheless, Azam, who tried to argue that 'there are no theoretically compelling reasons why this type of policy reform (SAPs) should especially harm the poor' (Azam 1994: 111), was forced by his own analysis to conclude that 'the devaluation-induced inflation will normally mainly affect the urban poor and the poorest rural workers who do not own any land or real assets' (ibid.: 103).

We have noted earlier, that the question of poverty alleviation was not central to the 'first wave' of adjusters, and that the social dimension to adjustment did not become a major concern in the adjustment agenda until after the publication of UNICEF's report, *Adjustment with a Human Face*, in 1987. It was felt that adjustment policies would not lead to major disruptions, and where this did occur, it would be short term in duration, and would persist only because of the rigidity of the pre-existing economic and political programmes. In Ghana where adjustment has been heralded as a success by stimulating growth at an annual average of 6 per cent between 1984 and 1989, poverty alleviation has not been achieved (Seshamani 1994: 114). In addition, 'inflation was confined to an average of less than 23 per cent compared to the three digit figures for the preceding years'.[3] Seshamani paints a similar picture for Uganda: despite a growth rate of 5 per cent each year since 1992, and a low monthly rate of inflation (0.3 per cent), 'Ugandans are worse-off than they were 20 years ago' (ibid.: 114).

Adjustment policies in the Third World impact on the poor and public workers such as teachers, whose salaries (in the case of Africa) fell by 33 per cent (Daddieh 1995: 41), through the inflationary pressure triggered off by: retrenchment, the 'necessary evil' of the Economic Recovery Programme (Jonah 1989: 141); and the adjustment and stabilisation programmes which follow – devaluation, leading to increased cost of imported items; decontrolling of prices; removal of subsidies from essential commodities; privatisation and the institutionalisation of cost-recovery programmes for social welfare facilities such as health and education.

In the case of Venezuela, a deeply recessive programme started in 1988 produced a fall in the GDP of 8.57 per cent in 1989 'as well as a record inflation – increasing from 29.48 per cent in 1988 to 84.46 per cent in 1989' (Lander 1996: 53). Even in 'flagship cases' such as Ghana and Venezuela, where successes have been registered, they have been short-lived, and many of these accomplishments have been reversed (ibid.; Brydon and Legge 1995). Lander has noted that in Latin America adjustment has had 'de-industrialising effects and accelerated the deterioration of the agricultural sector as the financial and service sectors grew' (Lander 1996: 58). In many of these countries, particularly those with a significant industrial base, industries tried to survive by shifting the cost of adjustment to the workers through labour-cost reduction, downsizing of the labour force, reduction of real wages, flexible labour market, intensification of work and alteration of the skill-mix (ibid.). In Latin America the minimum wage fell by a quarter in the 'lost decade' of the 1980s, and average earnings in the informal sector, where victims of SAPs sought refuge, fell by 42 per cent (Munck 1994: 91).

The net results of these policies are an enormous rise in the cost of living. These policies demand sacrifices and increased suffering in the short run (Jonah 1989), particularly for those who are dependent on public sector employment or on fixed incomes, thus creating what Seshamani has called 'the new poor'. This group in countries such as Sierra Leone, Ghana and Nigeria includes thousands of middle-class professionals such as teachers, health workers and civil servants who go for months without wages or salaries and whose jobs sometimes vanished over night (Zack-Williams 1997). These are the very people who are expected to lead civil society as well as constituting the backbone of civil society, the harbinger of democracy (see chapter 5). In the case of Sierra Leone at least, these excluded intellectuals have turned out to be quite destructive (Richards 1996; Zack-Williams 1998).

Furthermore, declining state investment and reduced institutional capacity, particularly in the social welfare sector, have resulted in hardship for those social groups dependent on the state for employment and other services. Cuts in public expenditure, including education and social services, have been marked by collapsing infrastructure and falling enrolment of students at all levels, especially at secondary schools (Lugalla 1993). Lugalla has argued that in the case of Tanzania, the transformation of the social sector which was seen as the success story of Nyerere's reforming period, was now being undone by the austerity of SAPs, as expenditure on education was slashed from 11.7 per cent of government current expenditure in 1980/81, to 4.8 per cent in 1988/89 (ibid.: 198). The shortage in capital expenditure on education of SAPs was also marked. In addition, Lugalla observed that the primary sector was most severely affected as this fell from 28.3 per cent to 9 per cent as a share of total development expenditure. This posed a threat to the government policy of providing basic education to children of primary school age.

Tanzania was not alone in reducing social expenditure – Sierra Leone (Zack-Williams 1990), Zimbabwe and Uganda experienced a similar contraction (Engberg-Pedersen *et al.* 1996). Not surprisingly, in many adjusting African countries classes are over-crowded, and children have to provide their own furniture, such as chairs and desks for school (for those who do not want to sit on the floor), as well as their equipment such as exercise books and writing materials. Other sub-sectors, such as secondary, tertiary, further and higher, have equally been affected by adjustment. Here too, salaries fell by a third (Daddieh 1995), as budget priorities moved from a highly subsidised system to one based on private funding and cost recovery. These moves led to crumbling infrastructure and support services, as well as widespread unemployment. Those who remained in this sector soon realised that they need 'to take second and third jobs to make ends meet, (as) class times become flexible and discipline is lax' (ibid.: 41).

In recent years (post 1990) there has been a shift in the Bank's priority. Emphasis is now on the need to prioritise basic education, 'since it is intrinsic to development in the widest sense; empowering people, especially the poor, with basic cognitive skills is the surest way to render them as self-reliant citizens' (World Bank 1989: 77). However, as Engberg-Pedersen *et al.* have noted: 'for Africa as a whole the general trend is one of declining per capita expenditures within the social sector during both the 1980s and the 1990s' (1996: 55), and both the economic crisis and SAPs have fostered declining institutional capacity to deal with the problem of social sector reform.

Most African countries embarked upon adjustment with falling nutritional standards and deteriorating social services.[4] In the case of Ghana, the institutionalisation of cost-recovery programmes, whilst it did not seem to produce immediate improvement in health care provisions, resulted in 'falling attendance at government medical facilities', from 10.7 million in 1971 to 4.7 million in 1983, and 1.57 million in 1991 (Engberg-Pedersen *et al.* 1996: 199). As Hutchful has pointed out (Engberg-Pedersen *et al.* 1996), this is not to suggest that patients have no recourse to other medical help. In many cases, we witnessed the emergence of social involution, as people (low-income as well as previously better off households) resorted to traditional healers and self-medication, as a single monthly visit to a medical practitioner could cost up to 15 per cent of monthly incomes (ibid.: 199). Two important features characterise health expenditure in Africa in the 1980s and 1990s: the first is that public expenditure on health remained very small at only $7 per capita in the mid-1980s compared to $26 and $75 per capita for Latin America and Asia respectively. Also the absolute and relative expenditure on health fell consistently during the two decades (Logan 1995: 58).

Pointing to the harsh socio-economic conditions brought about by SAPs, particularly on vulnerable groups, Logan has questioned whether 'the market mechanism can spread its influence beyond the urban areas to facilitate modern health care delivery to rural communities and the poor' (ibid.: 61). In casting doubt on the suitability of the market to distribute health care to Africa's poor, a highly marginalised group, who lack purchasing power, Logan concluded that the market mechanism has the potential to intensify social inequalities.

Locating winners and losers

One early goal of the policy of adjustment was to create 'a short, sharp shock treatment for African economies' (Ravenhill 1993: 21) in order to rectify macro-economic imbalances. Like its penal counterpart, the 'three S treatment' always produces losers, without necessarily guaranteeing winners. Furthermore, since one of the major goals of SAPs is to effect a major change in the urban–rural terms of trade in favour of rural producers, this is bound to adversely affect urban dwellers. For example, the urban wage index showed a marked fall over the period for all adjusting African economies. In Uganda, the index of real wages fell from 100 in 1972 to 9 in late 1984. Zaire did not fare any better with the purchasing power of the minimum wage in 1982 being only 3 per cent of the 1970 level (ibid.: 36). The minority of rural inhabitants who are involved in producing tradables may gain from such a policy, but the contraction of the state's responsibility has posed tremendous problems for urban dwellers. In some African countries such as Nigeria, it has been estimated that the government controls at least 60 per cent of employment in the modern sector, and that up to one million workers were made redundant as a direct result of SAP between 1984 and 1989 (Fashoyin 1994).[5] Irregular salaries and electricity supplies, shortages of essential commodities such as fuel oil and staples were all features of the adjustment era. Urban wage earners, including the petty bourgeoisie or former 'privilege salariat' (as Osaghae (1995) has labelled this group), are among the losers in the economic decline triggered off by policies of adjustment and stabilisation.

By creating losers and winners, based on the degree of proximity to the tradable sector, SAPs have tended to intensify ethnic and regional conflicts, as various ethnic groups strive for an ever-dwindling share of the national cake. In Nigeria's case, Osaghae has distinguished ethnicity in the pre-SAP period, which he described as mainly 'public sector ethnicity', the result of an expanding bureaucracy in a period of boom. It was also a phenomenon of the urban areas. However, in the era of SAP ethnicity has become a more significant feature in both rural and urban areas. As Osaghae has observed, the intensification of ethnic tensions and conflicts is dependent on the extent to which SAP reinforced existing social, economic and political inequalities, or created new ones (Osaghae 1995). The growth in ethnicity is reflected in the rise of ethnic corporatism in Sierra Leone in the height of the adjustment period during the Momoh era (Zack-Williams 1991). This period saw the rise to political prominence of the *Ekutay*, an ethnic cabal around which decisions of state were taken, particularly those relating to promotions and state contracts. This system of governance was used as a way out of a crisis of legitimacy triggered by the SAP, and the President's inability to build a strong base of support within the ruling All Peoples' Congress. In the context of Sierra Leone, the misery of SAP and the neglect of certain regions were major factors in the challenge to the state posed by a social movement, the Revolutionary United Front, since 1991. Osaghae too has pointed out that diminishing resources and the deteriorating standard of living, reinforced by inequalities, were major factors in the struggle of the Ogoni people against the Nigerian state.

SAPs and the gender gap

Though macro-economic policies are always shrouded under seemingly gender neutral terms, Diane Elson has pointed out that this apparent neutrality disguises a deeper gender bias (Elson 1989: 57). She observes that:

> An important source of male bias in the design of structural adjustment programmes has come from the assumption that the household may be treated as a unity, and that there is therefore no need to examine intra-household relationships between men and women in the analysis of the process of adjustment.
>
> (Elson 1995: 211)

Engberg-Pedersen *et al.* have argued that despite the growing awareness of the social aspects of adjustment, 'women have been targeted (only) as part of wider groups' (1996: 5–7). Sparr (1994) sees this bias as an inherent feature of the neo-classical quest for a value-neutral science, which is premised on a positivist epistemology. Given the socio-economic position of women in Africa and in many Third World societies – as producers and carers of the young, as agricultural producers, as domestic workers, preparing food, looking after the old, the infirm and the sick, as well as managing household budgets – the general condition of the economy has greater social significance for women than for men. This is particularly important, since much of women's work goes unremunerated.

As Elson has warned, structural adjustment may have a catastrophic effect if policy-makers continue to assume an elastic capacity to cope with the socio-economic demands of adjustment:

> women's unpaid labour is not infinitely elastic – a breaking point may be reached, and women's capacity to reproduce and maintain human resources may collapse. Even if the breaking point is not reached, the success of the macro-economic policy in achieving its goals may be won at the cost of a longer and harder working day for women. The cost will be invisible to the macro-economic policy makers because it is unpaid time. But the cost will be revealed in statistics on the health and nutritional status of women.
>
> (Elson 1989: 58)

In her work on Central America, Florence Babb has observed that in Nicaragua, women were functioning as a shock absorber for adjustment measures by absorbing the shock through their paid work in and outside of the home, since women's unpaid and paid work are highly inter-linked (Babb 1996). Similarly, Elson draws attention to the fact that the dual burdens of unpaid labour – in social reproduction, on the one hand, and paid and unpaid labour in the production of goods and services on the other – do not qualify women to compete in the market on equal terms with men. Adjustment has a differential impact on different women, depending on their socio-economic status, geographical location and their position within the labour force.

Buying substitutes (i.e. the labour power of other women) for their unpaid work can compensate for the disadvantaged position of rich women in the market. Thus middle-class women may employ maids, servants and cleaners to ameliorate the demand on their time. According to Elson, this does not obliterate their disadvantaged position *vis-à-vis* men in the market, since access to the market is shaped by access to the public sector services. As she points out:

> For all but the well-off women, there is a complementarity between state provision of services required for human resource development, and the ability to make gains from participation in the market. For most women, the choice is not between dependence on the state and dependence on a man.
>
> (Elson 1989: 63)

The attenuated nature of public services under SAPs has impacted on women, this 'quiet army', who as grand-mothers, mothers, sisters and aunts now have to substitute for the dwindling social services (Sparr 1994). Data from Latin America and the Caribbean also show that economic restructuring is increasing the importance and visibility of women's contribution to the household economy as more women enter the labour force to supplement the household budget stemming from the loss of men's employment. The share of women in the labour force rose from 32 per cent in 1980 to 38 per cent in 1988 (Safa 1995: 33). In a number of these countries, such as Puerto Rico and the Dominican Republic, the demise of the traditional industries and the expansion of the labour-intensive export manufacturing and service sector have meant that male participation in the labour force has either declined or remained stagnant (ibid.).

Various studies have indicated that SAPs have tended to intensify women's household management problems as they increase their efforts to maintain the family by redoubling their remunerated and non-remunerated tasks (Elson 1989; Brydon and Legge 1995; Zack-Williams 1995b). Redundancy among men has meant loss of the husbands' earnings and fall in the purchasing power in the family. This has forced more women to look for work outside the family. In other words, women's responsibility has increased without the endowment of social power. Adjustment, and the crisis it is intended to remedy, have helped to restructure the role of women within the household, without overthrowing domestic patriarchy. Household income has seriously been eroded in real terms, yet, men expect 'service as usual': food, laundry and childcare. The strain of managing an ever-diminishing budget is the woman's, and this is a major source of domestic conflict. Whilst middle-class women can fight inflation and shortages through bulk buying, they have to deal with irregular electricity supplies, which may result in the destruction of all the food in a refrigerator.

Adjustment seeks to reduce the economic activities in which women are concentrated, the 'non-tradable' sector. Paradoxically, adjustment tends to intensify women's domestic activities by depriving them of job security and protection (Manuh 1994), by pushing them out of the public sector through privatisation (Hatem 1994), as well as impelling them to seek paid employment to supplement the dwindling family purchasing power. Thus we can argue that SAPs tend to intensify the oppression of women so that wage differentials between men and women have been worsened by structural adjustment (ibid.). As we will see below, the policy of treating the household as a unity, and one that is always headed by men, means that compensatory programmes designed to cushion the worst effects of adjustments do not ameliorate the position of women (Laurie 1997).

Coping with adjustment

So far, we have argued that attempts to utilise structural adjustment programmes to rectify macro-economic imbalances in African economies have had far-reaching effects on African economy and society. In what follows, we now want to look at the survival strategies of various vulnerable groups. In looking at how various groups survived adjustment, it is important to note that the coping strategies for individuals and groups will vary depending on the impact of the SAP, as well as the gender and social positions occupied by such individuals and groups. The erosion of living standards impelled many individuals to take defensive action in order to safeguard their living standards. This took the form of seeking additional income by engaging in multiple jobs, or what Mustapha has called 'multiple modes of earning', part of a much wider phenomenon of 'multiple modes of social livelihood' (Mustapha 1992).

Mustapha argues that the strength of this concept is that it goes beyond the static analysis of 'marginality' and the 'informal sector', and should be seen primarily as a mechanism for income generation, and that its functionality may include illegal activities. Thus a university professor who decides to run his private car as a taxi to supplement his meagre salary, without obtaining an excise licence clearance from the authorities, or who decides to sell lecture notes to students, is in breach of the law just as much as the ambulance driver who siphons petrol from his vehicle to sell to private operators. Both social actors are seeking to supplement household finances in a milieu of rapid financial change. In the depth of the crisis, in Sierra Leone, Nigeria and Ghana, many professionals ran their cars as taxis; government drivers, as well as those from the private sector, used opportune moments 'to earn extra cash' by running taxis. This reflects both the scarcity of fuel, as well as a dearth of public transport. The inability of the state to make regular remuneration to its employees forced many into 'moonlighting' (in Sierra Leone this is referred to as *Mammy Coker*, or *Dregg*). The latter term, as it relates to the survival strategy of young women, has some worrying connotations. *Dregg* in this sense could take the form of undertaking demeaning tasks, mendicancy, as well as casual sexual favours (Zack-Williams 1990). Mustapha also points to a similar tendency in Nigerian society.

In many cases, the formal sector now acts as a supplement (when salary is paid) to the remuneration from the informal, or parallel, economy. The latter is the locus of much multiple-mode survival strategies and is continuously being expanded as an ever-increasing number of urban dwellers (including a growing number from the middle class) move into this sector as a direct result of adjustment policies and growing impoverishment (Tevera 1995: 86).

The informal sector consists of three sub-sectors: petty trading, small-scale manufacturing activities and the service sector (Brand *et al.* 1993). In many African countries, the informal economy is almost autonomous of the state, and it remains unregulated, with low overhead costs. In most cases prices are not fixed, but are negotiated between buyer and seller through haggling. The price paid is often a reflection of the buyer's perceived ability to pay, judged in terms of social status. The attraction of this sector stems from its unregulated nature, which means that prices tend to be well below those of equivalent products and services in the formal economy. The size of the informal economy in Africa has been put at between 50 per cent and 70 per cent (Cornia *et al.* 1987). In the case of sub-Saharan Africa's most populous country, Nigeria, Mustapha has argued that the informal economy clearly predominates:

> Under SAP ... it is likely that a larger proportion of the working population is now dependent on informal sector activities, possibly pushing the percentage in most Nigerian cities to close to 70 per cent of the working population.
>
> (Mustapha 1992: 195)

Through the various SDA (Social Dimensions of Adjustment) programmes and the exigencies of coping with SAP, Africans are being sold the value of the entrepreneurial culture, and an ever-increasing number of civil servants, politicians and retrenched workers set up business ventures. People such as teachers and health workers, who could not afford the capital for a business venture, try to set up private classes and drug stores respectively. In a few cases the expertise and loans provided by the SDAs have been utilised, but in most cases these are free from such assistance. The irony is that some of the African elite whose mismanagement caused the crisis that hemmed their nations into adjustments have now set themselves up as business consultants to foreign non-governmental organisations (NGOs), donors and multinational corporations. The business culture is dominated by trading activities, and child labour plays a crucial role in this sector. Many parents who cannot afford school fees, withdraw their children, usually girls (Lugalla 1993), from school in order to help bolster household incomes as hawkers.

One important mode of survival is 'the return to the land', which is in-built in some SDAs such as PAMSCAD, and the cultivation of 'any available plot of land'. The price inflation, particularly of food items, propels many professionals, and other groups with access to land, to grow cash crops and subsistence items. Others have turned to poultry, running piggeries and dairy cattle units (ibid.). It is important to note that though multiple-mode strategies are a feature of all social groups, for the poor it is the buffer between surviving and going under. On the other hand, 'for the professional classes ... the threat of survival is not that stark and dire. ... For these classes, multiple mode activities are seen essentially as means of containing, and possibly reversing the obvious slide in their living standards' (Mustapha 1992: 201).

In response to foreign exchange shortages, a small section of the professional class in a number of countries such as Sierra Leone, Ghana and Nigeria have now moved in to organise petty commodity production of artisans, craft workers, tailors and textile workers (Zack-Williams 1994). The products of artisans are organised by merchant capital (professional class), who sell them to cultural entrepreneurs operating in Europe and North America, who now assume the role of the colonial trading houses (ibid.). They purchase items of cultural significance to be sold to ethnically conscious consumers.

Finally, in the 1980s, many Third World people turned to religion as is reflected in the rise of spiritualist churches and secret cults, ethnic fraternity and sorority, and alumni associations. These religious organisations did not only act as the 'opium of the people', providing solace and comfort to troubled minds, but they provide an avenue for accumulation as many of them were run along business lines.

SAPs and poverty alleviation

In this section we want to look briefly at some of the social dimensions of adjustment programmes (SDAs) that have been instituted by the Ghanaian, Nigerian, Sierra Leonean and Peruvian governments. The point is to highlight some of the assumptions and weaknesses of these projects which tend to marginalise vulnerable groups, in particular, women. As noted earlier, the publication of UNICEF's *Adjustment with a Human Face* marked the beginnings of pressure on the international financial institutions to start addressing the human dimension of adjustment. The report noted that after three decades of advancement, the welfare of children in many parts of the world started to falter, particularly those in Africa, Latin America and the Middle East. It pointed to the direct link between economic decline in these regions, and the worsening fate of vulnerable groups such as women and children, in particular the under fives.

The report called *Adjustment with a Human Face* suggested a new 'range of economic and other policy measures' to be integrated into the policy programmes for each national government. This approach, it was felt, would allow Third World nations to regain the 'momentum of growth and development and to tackle the rising problem of debt, poverty, and social strain' (Cornia *et al.* 1987: 3) through a people-sensitive approach to adjustment. It noted that nutrition and education seemed to be the areas where deterioration had been most pronounced and as such these are in urgent need of attention in order to avoid human and economic miseries (ibid.: 6 and chapter 16). In this way, the report argued that adjustment with a human face should be seen as part of the process of poverty alleviation. The debilitating effects of both the crisis which called forth adjustment and the cure itself posed a threat to 'the mental and physical capacity of the future labour force' (ibid.: 288). The report noted that the type of adjustments which these countries have implemented are a contributory element to the deteriorating conditions now being experienced by the poor of these countries.

The report pointed to two important factors, which have prevented sustained growth, thereby undermining child welfare: the deflationary nature of most adjustment programmes, which reduce demand, depress employment and real incomes, and the direct negative effects of some macro-economic policies, particularly those relating to cost-recovery programmes, abolition of subsidies on essential commodities, cuts in social expenditure and the liberalisation of exchange controls on vulnerable groups, such as the poor and those on fixed incomes. The negative effects are the product of four factors: (1) the short-term horizon of SAP; (2) the predominance of macro-economic policies over sectoral targeted ones aimed at ameliorating the condition of the poor; (3) insufficient finance, leading to demand contracting responses; (4) lack of serious interest among policy-makers on the question of the impact of such policies on income distribution.

The report highlighted the lessons from the experiences of ten adjusting countries. The first is that adjustment is necessary if countries are to adapt to a changing world environment. Though necessary, growth-oriented adjustment is not enough if vulnerable groups are to be protected in the short to medium term. The report was at pains to point out that most vulnerable groups could be protected during adjustment, even in the absence of growth, at least in the short term if targeted programmes are adopted. The report pointed to successful alternatives to adjustment such as those implemented by South Korea. Finally, the report noted that the protection of vulnerable groups does not only raise human welfare but it is also economically efficient, if only because worker productivity rises with nutrition.

In order to protect vulnerable groups the report called for:

- more expansionary (and by implication less deflationary) macro-economic policies in order to raise output, investment and human needs over the period of adjustment;
- greater utilisation of meso-policies in order to prioritise policies in favour of the poor;
- sectoral policies should be designed in order to restructure the productive sectors so as to raise the level of employment and income-generating activities;
- compensatory programmes to protect basic health and nutrition of the poor through public works programmes and nutritional interventions;
- finally, the monitoring of the human situation, especially living standards, health and nutrition of the poor.

The authors of the programme emphasised that these were realistic aims given the existing global situation, and that each of these five canons of adjustment with a human face has been successfully adopted by a number of countries.

One of the first results of the human face of adjustment was Ghana's Programme of Action to Mitigate the Social Costs of Adjustment (PAMSCAD). Similar projects were set up in Peru (Laurie 1997), Sierra Leone (Zack-Williams 1997), Ivory Coast, Gambia, Guinea, Mauritania and Senegal (Ravenhill 1993: 36). PAMSCAD is the best known of all the SDA programmes. It was instituted by the government of Ghana in 1987, following the economic recovery programme (ERP). The government adopted PAMSCAD because 'despite promising medium-to-long-term growth prospects, the economy (remained) characterised by widespread poverty and economic hardship' (PAMSCAD: 3). It was a short-term programme designed to deal with the problem of poverty by looking at the needs of vulnerable groups adversely affected by the ERP. These vulnerable groups included small farmers (mainly in the north of the country), low-income, unemployed and under-employed urban households, as well as retrenched workers in the public and private sectors. The setting up of PAMSCAD is an acknowledgement that: 'The ERP will not be able to alleviate the economic hardship of many of the poor and vulnerable groups in the short run ... and will exacerbate the economic problems of certain vulnerable groups' (ibid.: 2).

The total financial requirement for PAMSCAD was US$83.9 million, of which there was a foreign exchange component of US$37.6 million that was funded by aid provided by donors (ibid.: 4). A number of criteria were laid down in order to qualify for funding:

- A project must have a strong poverty focus specifically designed to benefit vulnerable groups.
- It must have high economic and social rate of return, and be cost-effective.
- Implementation of the project must be relatively easy and quick.
- Intervention must not create unsustainable future obligations for recurrent costs.
- Such interventions must not create distortions, nor conflict with the larger objectives of the ERP.
- Finally, in the area of redeployment and education, intervention should have politically marketable currency, and high visibility.

On a number of indices, it is doubtful if PAMSCAD can be described as a success story. First, one important vulnerable group, women, did not seem to have been central to the thinking of those who drew up the programme. Not surprisingly, very few women knew of the activities of PAMSCAD even in areas where officials had undertaken a project (Brydon and Legge 1995):

Amedzofe (a village in Volta Region) has been the recipient of PAMSCAD Community Development funding to build a latrine block. The block was 'inaugurated' with great ceremony by visiting PAMSCAD officials less than six months before we began work. Of those women in our Amedzofe sample, even so shortly after this grand opening ceremony, almost one third said that they knew the name only, not what it signified or what its purposes were.

(81)

This is not surprising since most of the women surveyed (77 per cent) said that they had never received loans from the bank or PAMSCAD officials, who were accused by the women of 'just making fun of us' (Brydon and Legge 1995: 72).

PAMSCAD, like other SDAs in Africa, suffered from the fact that there was over-dependence on external donors for funding projects. And like its Sierra Leonean counterpart, SAPA, PAMSCAD took a long time to get off the ground (Martin 1993: 162). In the case of the latter, its success was too strongly tied to the PNDC's (Provisional National Defence Council) quest for decentralisation at the district and sub-district levels. Failure of the decentralisation project had major implications for the activities of PAMSCAD because local planning was not co-ordinated. Finally, the need to be above board with the ERP meant that any programme which was seen to run counter to the aim of the adjustment programme was rejected, regardless of long-term sustainability.

In Latin America the SDA programmes have been more focused on employment generation and 'workfarism' as opposed to small-scale infrastructural projects. For example, as part of its SDA the Peruvian government embarked upon a programme of support and temporary incomes, mainly designed to cope with unemployment, by providing emergency work for unemployed men. As Laurie has pointed out, this programme was based on the belief that breadwinners were men who needed to earn a 'family wage'. As such PAIT (Programme of Support and Temporary Income) was riddled with 'male bias', in that it assumed the male wage is equivalent to the family wage and that female wages are secondary (Laurie 1997). The assumption was that women's labour was exclusive to the domestic sphere. In this way the Government ended up consolidating motherhood, by not constructing a programme that perceives women in their true public position as earning wages. Laurie warned:

> Notions of a male wage being a family wage and a female wage being secondary are influenced by such complex ideological legacies. The definition of welfare adopted by APRA (Government) saw women as members of families with a specific role to play; therefore the party conceptualised women's wages as additional to a household (male) salary. Women's PAIT wages became an economic help (*ayuda*) to the household rather than the main support (*apoyo*).
>
> (Laurie 1997: 703)

Laurie argued that both globalisation and the rise of neo-liberalism have tended to create extra competition for women in the wage sector, but policy-makers have given very little attention to this development. Thus when the local SDA was conceived in Peru, the assumption was that since unemployment was a problem for men, this programme was designed to serve the interests of unemployed men. Thus she observed:

> State assumptions about a demand for emergency work from unemployed men were largely based on the idea that men need to earn a 'family wage' ... in the case of PAIT male bias is seen in the fact that the Peruvian state assumed a 'family wage' would be earned by men who constituted heads of unified households and made household decisions for 'the good of the unit'.
>
> (Laurie 1997: 701)

Laurie noted that the government was surprised to see that PAIT was dominated by women who under the welfare-to-work logic were not constructed as wage earners, and as such were not part of the labour force. Thus as Laurie pointed out PAIT was feminised, as women moved into jobs that were designed for men. Once this feminisation of PAIT became public knowledge, the Peruvian state took measures 'aimed at cementing this change from productive work to the provision of welfare services' (ibid.: 704), thus confirming the socialised role of women as 'mothers receiving welfare rather than as employees receiving "on the job" training' (ibid.: 705).

Conclusion

In this chapter we have looked at the social consequences of SAP on various social groups in the Third World. We noted that one problem in locating these consequences for the population of the Third World is the difficulty of disaggregating the effects of SAP and those of the crisis that impelled these countries to seek assistance from the IFIs. We argued that one way forward was to look at the impact of specific policies on the population. We noted that prior to UNICEF's *Adjustment with a Human Face* the social dimension to adjustment was not a major item on the adjustment agenda. Following the publication of this report, several SDAs were set up in a number of adjusting countries. In looking at PAMSCAD in Ghana and PAIT in Peru, we drew attention to the fact that because these programmes are premised on a false assumption about the role played by women in the household, these programmes tend not to serve the needs of women. Indeed, in the case of PAMSCAD we noted that very few women knew of the existence of the project or its function. Similarly, PAIT which was designed to serve the men as wage earners within the household was quickly transformed into a welfare facility, once the Peruvian government realised that PAIT had been feminised. The high dependence on external funding was also a major problem for these projects. For example, SAPA in Sierra Leone did not take off for a long time due to lack of funds to start the projects. SAPs destroyed not just the urban poor, but, by impoverishing the middle-class professionals, we could argue that it destroyed the basis for genuine democracy.

Notes

1   Though D.E. Sahn, 'The Impact of Macroeconomic Adjustment on Incomes, Health and Nutrition: Sub-Saharan Africa in the 1980s' in G.A Cornia and G.K. Helleiner (eds) *From Adjustment to Development in Africa: Conflict, Controversy, Convergence, Consensus?* (London: Macmillan Press, 1994), has argued that 'an unweighted average of social sector spending among African countries during the 1980s indicates that health and education expenditures were on the rise during the 1980s' (275). The stagnation in per capita spending he puts down to the demographic upsurge.
2   Writers such as E.E. Osaghae have argued that 'the period of economic recession that led to SAP and that of SAP itself should be regarded as part and parcel of the same regime' (1995: 40).
3   *Programme of Action to Mitigate the Social Costs of Adjustment*, The Government of Ghana, November, 1987, p. 5.

4   E. Hutchful 'Ghana' in P. Engberg-Pedersen *et al.* (eds.) 1996: 197. He argued that between 1975 and 1982, real expenditure on education fell by two-thirds and that the health budget fell by almost 80 per cent.
5   Ibid. It has been estimated that by the late 1980s unemployment in Nigeria reached the three million mark, and three-quarters of these were people aged 25 and under (National Directorate of Employment, *Creating More Job Opportunities*, Federal Ministry of Labour and Productivity, Lagos, Nigeria, no date, p. 2).

5 Contested sovereignty and democratic contradictions

The political impacts of adjustment

Giles Mohan

Introduction

This chapter firmly rejects approaches to political analysis which privilege the state as the prime actor. This has two main implications. First, for many analysts in 'the North' there has been the relatively recent discovery that globalisation signals an erosion of state power. This reflects certain realities such as the increased mobility of capital, the powerful role played by TNCs in undermining national regulatory regimes, and the emergence of regional politico-economic groupings. However, it also reflects a privileged and blinkered Eurocentrism regarding hegemony and power in the world economy. For most countries undergoing SAPs which, as we have seen, were colonised at one time, the discovery of eroded sovereignty and autonomy is perverse. From their creation these states have been heavily influenced by the geo-political and geo-economic interests of the coloniser and/or superpowers. The second implication of overly state-led analysis concerns the linkages between the state and civil society. By focusing on the formal arrangements of the state, be they corporatist negotiations or electoral democracy, these analyses ignore the complex and important ways in which much of 'politics' is actually elaborated. These concerns overlap with discussions throughout this book, but in particular those found in chapters 4 and 6.

This chapter argues that structural adjustment represents a qualitative change in the relationship between lenders and nation states which has been driven largely by the power conferred on the lenders through debt. Related to this is the fact that structural adjustment has resulted in unprecedented effects upon politics and most of the policy process. All of these factors conspire to set up an irreconcilable tension between democratic participation and SAPs despite much rhetoric to the contrary. The chapter is divided into four sections. The first examines the general debate around the state and market since this sets the context in which the politics of adjustment takes place. In it I discuss the changing nature of this debate which has swung pendulum-like between advocating the state or the market and presenting the choice as either/or. The next two sections explore the policy process and the restructuring of the state involved in doing this. As such the sections look at policy formulation and implementation respectively although there is clearly blurring and feedback in this movement. Finally, the chapter discusses the more recent advocation by various institutions of 'good governance' which is a loosely defined term covering state capacity, democracy and civil society. In doing this I touch on the issues of participation and representation which takes us into the analysis of chapter 8.

The state–market debate and the neo-liberal backlash

While the introduction to this chapter stressed a decentring of statist analysis this does not necessarily imply that the state is no longer an important political referent. Indeed the national state level still plays a pivotal role in mediating the global and the local and it has been the focus of much attack under SAPs. However, we must not view the global, national and local as discrete political realms but conceive of them as overlapping and intermingling (Slater 1993).

It is no longer possible, if indeed it ever was, to view national political spaces as hermetically sealed so that it is equally misleading to talk of internal and external forces since no action is completely immune from other reciprocal actions (Ashley 1987; Bayart 1993). We begin this section by looking briefly at the modernisation school and its views of the neutral state, the neo-Marxist school who emphasised the dependent nature of the state and the subsequent need to de-link from the world economy. More recently state intervention has been attacked by neo-liberals who see it as an impediment to efficient markets. They take a zero-sum, either/or view of political economy which is simplistic and short-sighted.

Modernisation, dependency and Keynesian development economics

In this section I provide a very brief and generalised analysis of the theoretical alternatives which preceded neo-liberalism. These fall loosely into a Keynesian/ modernisation axis and a neo-Marxist camp. This is a partial account, but is included to emphasise the centrality accorded to state which the neo-liberal adjustment model sought to dismantle.

Leys (1996) argues that 'development', both as a normative group of theories and a set of practices, has been most closely associated with post-war Keynesian programmes. It was in this period that state intervention and market regulation were most widespread as policy-makers attempted to reconstruct war-shattered economies and prevent the cyclical fluctuations that had intensified geo-political rivalry (Preston 1996). Keynes argued that self-regulating markets could not exist and that unemployment, resource misuse and increasing monopolisation ran contrary to the credo of the *laissez-faire* equilibrium theorists (Hutton 1986).

The post-war period saw the conjuncture of a change in economic and political thinking, the growing hegemony of the USA (and the attendant superpower conflicts), and the emergence of newly independent states which mobilised around nationalist development. This conjuncture saw the application of Keynesian policy in Third World countries, refracted through the prism of growth theory and development economics (Preston 1996). These blueprints were implemented through centralised planning agencies so that the development plan became a symbol of progress and modernity (Conyers and Hills 1984).

Throughout the 1950s it became clear, especially through India's experience, that this positivist approach to development planning was not yielding the desired results. In particular, despite the most comprehensive plans, it was obvious that many states were too weak to enforce these policies, but more importantly, the people were unwilling to conform to these imported notions of 'change' and 'progress'. It was here that modernisation theory emerged as a means of redirecting the cultural logics of these societies. Such concerns were clearly tied to the Cold War agenda and through a mixture of Weberian and structural-functionalist social and political analysis these theorists argued that the problem lay with the 'traditional' values of the masses. It followed that a benevolent elite should be the catalyst for 'modern' values to be diffused through society.

The political modernisation school which emerged in the early 1960s was heavily influenced by Cold War geo-political concerns tied to a teleology which implicitly viewed Western (read American) society as the terminal point. In this model economic growth occurs in inevitable stages, but political institutions only arise after economic development has occurred. The state is regarded as a homogenous 'black box' whose role is simply to translate inputs into outputs (see the discussion in Randall and Theobald 1985; Hague *et al.* 1992). Given their Eurocentric structural-functionalist roots these models also assumed that every state necessarily desired a similar sort of development which negated the possibilities of alternatives or dissension.

The neo-Marxist response argued that dependency is inevitable and that the core–periphery situation is enabled and perpetuated by a 'comprador bourgeoisie' which benefits from under-development (Alavi 1972; Cardoso and Faletto 1979; Leys 1996). These interventions were crucial in highlighting the economic basis of politics and development. It also focused attention on power and imperialism and sees the state as a necessary factor in this process. In this sense these theories were a valuable riposte to the Eurocentric liberalism of modernisation theory and posited a working class/peasant revolution leading to state control as a possible solution to structural underdevelopment. However, like modernisation theory, the state's role is reduced to servicing the needs of international capital which greatly reduces the scope for analysing politics (Randall and Theobald 1985). Both the modernisation and neo-Marxist schools share structuralist origins which produce teleological and deterministic accounts of politics (Bayart 1991, 1993). This brings with it relatively totalising accounts of history which denigrate the importance of culture and human agency.

Clearly these approaches are very different in their ideological underpinnings and their attitudes towards the influence of the market. But what the Keynesian and Marxist views stress is the centrality of the state as either regulator/redistributor or comprador. Clearly modernisation theory is more positive about the benefits of capitalism such that the state's role is to facilitate this transition from a 'traditional' to 'modern' economy. So, in the mid-1970s, as we saw in chapter 1, the pendulum shifted away from the state as agent of development towards a view that it was precisely the state's misallocation of resources that was counter-productive. This ushered in what Toye (1993) referred to as the 'neo-liberal counter-revolution'.

The neo-liberal counter-revolution

In many ways the neo-liberal counter-revolution sees a return to the assumptions of the modernisation school. In valorising markets it had to delegitimise the state such that it set up a simplistic opposition between states and markets. For example, in a polemical essay Milton and Rose Friedman exclaims 'voluntary exchange is a necessary condition for both prosperity and freedom ... the government is the major source of interference with a free market system' (Friedman and Friedman 1980: 11 and 17). Much of the analysis which supports these claims is based around reductive rational choice theory (RCT).

The rational choice school of politics goes back to the Cold War era (Downs 1957; Olson 1965) but it is only now becoming prevalent in lender thinking, especially that of the World Bank (Leys 1996). In its earlier forms, RCT contributed to the positivist-functionalist social sciences that dominated the US academy. It is with the spreading of neo-liberalism through debt leverage that it has found its way into development theory which has always been amenable to simplistic and mechanistic policy recommendations (Toye 1993; Escobar 1995). This so-called New Political Economy (NPE) comprises a number of related branches which share many of the rationalising assumptions of neo-classical economics but builds upon them in critical ways. These developments centre on extending the neo-classical project of understanding market equilibrium to exploring the ways in which market failure occurs.

The spread of RCT perspectives into development theory and policy has been a recent phenomenon. The World Bank increasingly cites the works of Robert Bates and Douglass North, even though there are relatively few empirical applications of these theories in a development context (Nabli and Nugent 1989). In evaluating the influence of these ideas it is useful to see them as a field of discursivity in which power and knowledge and theory and practice are intimately bound (Escobar 1995; Crush 1995). In doing this it is not enough to dismiss the ideas as 'rhetoric', as Toye (1993) does, but to situate their significance within the wider political-economic context of neo-liberalism wherein these ideas both shape and reflect empirical events.

The broader neo-liberal agenda revolves around a profoundly cynical and pessimistic view of the state in which state actors are only in politics for personal gain which means that 'correct' decision-making does not prevail, the obvious corollary being that the unfettered market will deliver efficient and equitable results. As Manor (1995) notes, the lenders 'assume such a radical opposition between states and markets' (311). In this sense NPE is being applied very flexibly since the original analyses were less about 'state versus market' and more about institutional processes within markets and between markets and the state. Hence the underlying tendency is to create the 'narrative of capital' rather than the 'narrative of community' (Williams and Young 1994). This can be seen in Kreuger's (1974) analysis of the rent-seeking state which sets in train a vicious cycle of pessimism towards both the market and the state because neither act 'efficiently'. A decade later Lal (1983) attacks 'dirigiste dogma' where 'serious distortions are due not to the inherent imperfections of the market mechanism but to irrational government interventions' (quoted in Corbridge 1995: 63).

More recently theorists from very different backgrounds have begun to endorse these institutionalist ideas. One of the most pervasive in analysing African politics is the ·statist school which derives from the work of Theda Skocpol (Evans *et al.* 1985). Adherents (Hyden 1983; Forrest 1988; Rothchild and Chanzan 1992) see the state as operating 'autonomously' and they stress 'the importance of the state and state actions in grasping the roots of the political and economic crises' (Rothchild and Chanzan 1992: 20). The attraction of this approach for an analysis of Africa is based on the existence of a self-seeking political class which seemingly acts 'autonomously'; so much so that Hyden (1983) sees the state as a balloon suspended above society. These views are not confined to this school, with a more left-oriented scholar claiming that Africa's developmental crisis is 'primarily ... a crisis of institutions' (Davidson 1992: 10). The implication of all these approaches is that both state and civil society are independent, internally consistent and mutually exclusive and one is unilaterally determining in the last instance (Jessop 1990). Recently this institutionalism has been applied to more specific policy areas such as decentralisation (Rondinelli *et al.* 1989) and corruption clean-ups (Mbaku 1996) while the thrust of governance programmes, as we shall see in the fourth section, is to create an 'enabling environment' for private investment (World Bank 1992b; Chalker 1993).

On the other side of the neo-liberal political armoury is the promotion of civil society which, as we will see below, has become an important element of the governance agenda. As Williams and Young (1994: 87) note, 'The Bank's promotion of civil society is linked to its promotion of accountability, legitimacy, transparency and participation as it is these factors which empower civil society and reduce the power of the state.' In this sense power is viewed in terms of a zero-sum game because it treats power as bounded nationally but as the preceding discussion showed this is an ideological ploy. Implicit in these formulations is the view that civil society is a bulwark against the over-bearing state and provides a check on excessive use of state power (Meiksins Wood 1995). Again there is a conflation of the 'economic' and 'political' by ignoring economic agents as political actors and thereby presenting civil society as a purely 'political' sphere (ibid.).

These debates help us explain the misconstrued notion of market–state polarisation and the idea that state and civil society hardly interact and reveal the ideologically charged nature of their prescriptions. Here I follow Beckman (1993) who argues that the state represents the only viable focus of resistance to world market subordination so that efforts to promote civil society over the state are yet another means of facilitating the deepening of capitalist relations. He says: 'The "liberation of civil society" plays a vital role in the struggle to legitimate the shift in the balance of forces, both internally and globally, and to de-legitimise resistance and contending options' (Beckman 1993: 22). In this light the apparently 'radical' emphasis on empowering civil society is part of the broader (neo)liberal project and it is erroneous to see the two political spheres as independent since 'the state plays a central role in the construction of civil society ... and ... Both state and civil society are formed in the process of contestation' (21 and 29), the corollary being that any attempt to use the state effectively will not be premised on donor-funded support of civil society which aims to magically develop it as an 'autonomous' sphere.

This emphasis on atomisation and civil society has further implications. The neo-liberal assumption of the natural efficiency of the market means that social justice is not incorporated into its democratic vision (Gills and Rocamora 1992). According to Sklar (1996), neither Hayek nor Fukuyama mentions social justice in their grand schemas. This is different from the modernisation theorists who did link economic growth with some measure of redistribution. This reflects the underlying assertion that the market delivers the best justice and vindicates the non-action of trickle down. This reverses the basic needs approach of the 1970s despite the rather belated and half-hearted social costs of adjustment programmes which we discussed in the previous chapter (Cornia *et al.* 1987; Gibbon 1992).

This section has examined the evolution of political theories of development. The latest manifestation under the adjustment era has been to valorise the market over the state and reject seriously those policies which aim to redistribute wealth to poorer sections of society. We have seen in chapter 4 how the various social costs of adjustment programmes were a weak attempt to remedy this. Later in this chapter we look at how, in the face of mounting difficulties in implementing SAPs, the donors began to push even harder for political reforms under the veil of simplistic 'governance' initiatives. Having examined the theoretical underpinnings of the politics of adjustment we need to look at what actually happens in practice when SAPs are initiated in the Third World. This will be covered in the next two sections.

Contested autonomy: the politics of policy formulation

A key issue which needs to be addressed is the politics of SAP negotiation and formulation. As Held (1989: 196) observes: 'It has to be borne in mind that IMF intervention is routinely at the request of governmental authorities or particular political factions within a state and, therefore, cannot straightforwardly be interpreted as a threat to sovereignty.' Accordingly, it is not enough to simply attribute blame to the IFIs and see all policy as externally determined because the loans that underpin adjustment programmes were requested by governments. However, there has been criticism that SAPs have been applied in a doctrinaire and coercive way which leaves little room for manoeuvre. It is this tendency that prompted Bush and Szeftel (1995) to describe the BWIs as the 'new Stalinists'. This section attempts to unpick the complex processes of domestic 'consensus' for neo-liberalism and the processes of negotiation between adjusting governments and the BWIs. We need to examine what factors precipitated the move towards adjustment. What role did different class fractions within the countries concerned play in pushing towards adjustment? This must focus on concrete cases of the debates which preceded the SAPs and the ways that adjustment entered the formal political agenda and was transformed by leaders. We will examine these in case studies from West Africa.

The politics of somersaults and U-turns

Although it is dangerous to generalise about the experiences of these countries they tend to involve a combination of processes whereby the leadership had to formulate and internalise a neo-liberal 'consensus' which often necessitated removing opposition to adjustment. In particular we see the internalisation of neo-liberal ideology and the flexible use of essentially liberal political concepts such as populism, participation and the self/family/community nexus. This relates back to the end of chapter 1 where we discussed the relationships between globalisation as discourse, but in contrast to those analysts who begin and end with the 'text' (O Tuathail 1996) we need to embed discourse within concrete institutional and social settings. For example, the neo-liberal consensus in Chile was fought through the theoretical debates of the so-called 'Chicago Boys' who were trained in economics in Chicago and returned to 'apply' their neat models to the real world of the Chilean economy. General Pinochet's initial scepticism towards the neo-liberal model was soon appeased once the Chicago Boys 'flew in Milton Friedman and Arnold Harberger for a high-profile lobbying effort' (Green 1995: 25). So the interplay of intellectuals and policies is mediated by various institutions and counters the tendency to view SAPs as imposed on unwitting and helpless governments.

In the formal political arena the general acceptance of harsh austerity programmes in Latin America is problematic and in Latin America no party would be or has been elected on a strongly neo-liberal platform, because of the effects of the short-term 'stabilisation phase' on poverty and social dislocation which corresponds to the electoral cycle (Green 1995). Instead we see regimes campaigning on either anti-neo-liberal platforms or, as we shall see below, around highly flexible populist political discourses which appeal to a broad section of society without really conceding obvious policy resources. Green observes 'the men who subsequently became the darlings of the neo-liberals won the presidency by campaigning on an anti-neo-liberal platform, then performing a policy somersault on coming to power' (1995: 157). We can see this process demonstrated in the Ghana case study below.

An important element of the consensus-building process is 'where there is articulate opposition to policy reform, as there usually is, that opposition must be overcome somehow if implementation is to take place' (Mosley *et al.* 1991:

287). They refer to formal political opposition either within the ruling party or from opposition parties. On the other hand popular opposition to neo-liberalism was much less organised. As Green (1995: 158) observes in Latin America: 'Resignation to the inevitability of adjustment hardly constitutes real support' and he shows how popular opposition was muted largely because of the preceding years of hyper-inflation and economic trauma. So while SAPs were clearly not a safe electoral ticket, any promise of some form of economic stability would at least not stir up opposition and may, in certain cases, elicit cautious support.

This is not to suggest that these programmes were not contested. The potential for social disruption meant that leaders such as Siaka Stevens of Sierra Leone and Kenneth Kaunda of Zambia were not prepared to go the whole hog with SAPs. In many cases adjustment was only implemented reluctantly (Olukoshi 1994a; Lander 1996). For example, in the mid-1980s, the military government of General Babaginda of Nigeria called for a public debate in part to determine the legitimacy of the programme, as well as to foster local owner-ship of SAPs. As Van de Walle has noted such a debate can give respite to a beleaguered regime as well as a sense of ownership of these policies:

> it [public debate] can provide the government with some temporary breathing room as well as educating the public about the parameters of the crisis and the choices available to the nation.
>
> (Van de Walle 1993: 92)

Babaginda broke off negotiations with the Fund in September 1985, though he later implemented essentially the same reform programme, this time with little opposition. He argued in a subsequent speech that the *raison d'être* for the debate was to show that there was 'no painless way of correcting any economic structure that has been long profoundly distorted' (Umoden 1992: 453).

Ghana is another interesting case study of regime changes in and around the adoption of neo-liberalism. The country was held up as a shining example of a successful SAP, but in the early 1980s the government was very anti-neo-liberal. The official mood of the time can be summed up in a quote from President Rawlings: 'We need time to reduce our need of the IMF, by seeking alternative ways of production. Small is good, at least for us' (*Observer*, 13 Feb. 1983, p. 11, quoted in Rothchild and Gyimah-Boadi 1988: 254). The irony of this is that Kweis Botchwey and Kojo Tsikata had been in negotiations with the IMF for some time and agreed on a package of reforms amounting, for all intents and purposes, to a SAP. Bing (1984) describes the transition to neo-liberalism as one of the biggest 'U-turns' in post-independence African history. Broadly speaking there were ideological differences within the Provisional National Defence Council (PNDC) that were not apparent at the time of the coup in 1981 when they came to power. The so-called right was not ultra-conservative, but essentially bourgeois and therefore opposed to popular government. They believed in free enterprise and only limited state regulation. The left was composed of a relatively broad spectrum of groups and individuals. The more moderate elements such as Kojo Tsikata and Kwesi Botchwey believed in some foreign assistance to promote socialism while a group of younger more radical elements were vehemently neo-Marxist and believed in cutting all ties with the West. The left was basically opposed to a number of common concerns such as corruption, injustice and neo-colonial policies, but lacked a common ideological viewpoint. The radical left within the PNDC had the upper hand in the first year of its existence. The right were better organised politically, but the left were reassured by the PNDC's anti-colonial stance.

In late November 1982 there was a coup attempt from the left which brought to the fore the internal conflicts of the PNDC. In the aftermath, the radical National Defence Committee was dismissed for alleged infiltration by 'counter-revolutionaries', while key actors were deported. As Boahen (1989) points out these dismissals and resignations effectively eliminated the more extreme elements of both the left and right within the PNDC and therefore allowed the moderates such as Botchwey and Tsikata to dominate political decision-making within government.

Hence, the consensus which the PNDC reached was not so surprising as one would have imagined in December 1981. The acceptance of the ERP then was born out of extreme economic hardship and a political willingness to implement it once the radicals had gone. It was the objective economic conditions and not so much the avowed political leanings of the government which were crucial, thereby reinforcing Ray's contention that: 'The issue of the IMF loans is not whether they are acceptable to socialist governments ... but rather at what cost' (Ray 1986: 138). Clearly, the PNDC, while questionably 'socialist', felt that the costs were worth the cure. In fact: 'The task of securing financing for the participatory approach might well have been more difficult than the technical approach' (Bing 1984: 102). This shows that IMF conditionality is not consequent upon the loan but in fact precedes it.

Conditionality and policy formulation

Although not a neat linear political process, once a government has decided to seek assistance from the BWIs for a SAP the bargaining process opens up. In contrast to the more structuralist accounts (Hayter 1971; Hayter and Watson 1985) which deny recipient governments any autonomy, Mosley *et al.* (1991) are more charitable on sovereignty, conditionality and the bargaining process. These authors assert that 'conditionality as practised by the World Bank is a *game* in the sense of being a relationship in which the two parties have (at least partly) opposed interests which they pursue by taking note of each other's likely behaviour, and in which the outcome depends on the strategies pursued by each party' (ibid.: 67–68, original emphasis). This 'game' is divided into three phases: initial negotiations leading to action items, implementation, and performance monitoring leading to possible further funding. The first and third phases are the bargaining stages where the two sides contest amounts of money and the conditions attached to them. Bargaining can, and does, break down where one party is unsatisfied with the terms and conditions.

An interesting twist to the bargaining process is that leverage is not always in favour of the BWIs. During the Cold War, many developing countries could play the superpowers off against each other and threaten to turn towards the other side if financial and military support was not forthcoming. It has been argued that with the collapse of the communist alternative, there is no counter-weight to the 'Washington Consensus'. However, SAPs have elicited so much popular and intellectual opposition that the BWIs are well aware that they need success stories in order to justify continued support and involvement. In Africa in the mid-1980s, Ghana was always the so-called 'front runner' while in the mid-1990s it was Uganda which was the darling of the financial institutions. In such, exceptional, cases these countries have greater bargaining power *vis-à-vis* the lenders. In Uganda, the defence issue has been a continual sticking point with the government increasing military expenditure from 13 per cent of the budget in 1991 to 20 per cent in 1995 in order to control incursions from Ugandan rebels in Sudan. President Museveni has been able to finance this mili-tary expenditure without losing the support of the BWIs although relations have been extremely strained at times (Channel 4 1998).

This shows that while this process is clearly not a simple imposition of poli-cies there is little space for autonomous policy actions and the neo-liberal cure tends to be similar across countries. However, we must avoid the tendency to simply say that all SAPs are the same and their effects equally positive/negative. As we have stressed throughout this book, only by exploring differences and similarities can we build up a useful picture of SAP intervention and its effects.

In concluding this section two important considerations need adding. The first concerns the role that SAP lending plays in releasing further funding from bilateral donors and private banks. In effect, the intervention by the IFIs acts as a rubber stamp for other lenders (Hayter and Watson 1985). As a World Bank memorandum stated:

> We expect that there will be opportunities to co-finance our structural adjustment loans with both official and private sources of finance but, more important in terms of volume, we expect that sound programmes of struc-tural adjustment, acceptable to the Bank as a basis for lending, will encourage bilateral aid agencies to expand lending for this purpose and complement and sustain lending from commercial banks for general purposes.
>
> (cited in Hayter and Watson 1985: 101)

The structural adjustment loans and the reform programme they initiate signal to other lenders a willingness to co-operate and accept 'rational' economic plan-ning. This reflects Agnew and Corbridge's (1995) claim that these international organisations effectively 'discipline' indebted countries.

The second, and perhaps most important, consideration is not so much who was involved in policy formulation but who was omitted. We have seen that neo-liberal policies are requested by recipient regimes who show a 'pronounced receptivity to external advice' (Haiti Country Program Paper 1983 cited in Hayter and Watson 1985: 109). As we have seen, SAPs are premised upon models which focus so exclusively on 'prices' that other important 'non-economic' factors are ignored. This non-consultation clearly has serious implications for processes detailed in other chapters where a variety of social institutions were ignored so that many of the impacts of SAPs were unseen and unpriced. For example, the gender and environmental dimensions were relatively underplayed so that subsequent policy outcomes, as we see in chapters 4 and 6 respectively, were damaging to women and the environment.

The paradoxes of implementing adjustment

During the implementation of SAPs profound political changes take place, although many are not obvious. It is, in part, this stealthy encroachment of political conditions, and the responses they elicit, which inspired Duncan Green (1995) to title his book on SAPs in Latin America *The Silent Revolution*. In chapter 7 he warns us that there has been a 'range of political models and processes that have accompanied the silent revolution' (Green 1995: 154). As we keep stressing we need to be wary of over-generalising about the effects of SAPs although there are some common processes. To demonstrate these I have combined some overviews with specific case studies.

Central to this section are the issues of authoritarianism and the contradictions involved in creating markets in various social arenas. This section examines the record of neo-liberal political change and focuses upon a number of key political sites. We saw that the IFIs fail to acknowledge their own political position in releasing national markets. This process in turn has serious implications for ideological struggles since acceptance of the neo-liberal package is predicated, as we saw, upon an erasure of competing political visions. As Frederick Chiluba, the Zambian trade union leader and later President, asked rhetorically: 'Where on earth has the IMF financed socialism?' (cited in Callaghy 1990: 297).

One particularly interesting phenomenon of recent years, which very effectively illustrates the relationship between neo-liberalism and regional political culture, comes from the rebirth of populist politics in a new neo-liberal guise (Weyland 1996). By the 1980s, Latin American populism was thought of as having been consigned to the history books, as the region's political leaders gradually embraced the need for neo-liberal economic reform and liberal democratic forms of political organisation became dominant. In this context, populism was depicted as the very antithesis of the new neo-liberal order through its association with 'irresponsible' state spending and inefficient and corrupt state-led economic strategies. If contemporary populist politics were referred to at all then they were 'equated with a lower-class backlash against the austerity, inequalities and market insecurities attendant on neoliberalism' (Roberts 1995: 83) rather than as part of the wider strategies of the region's political classes.

Any analysis of the recent electoral history of the region reveals, however, a much greater perseverance of populist rhetoric and a more general reluctance on the part of politicians to embrace the neo-liberal position wholeheartedly. None the less, this has most often been interpreted as part of a 'bait and switch' electoral strategy whereby populist campaign rhetoric against the austerity and pain of neo-liberal restructuring has been exchanged for neo-liberal economic policies once power had been attained (Drake 1991). Whilst certainly of some validity, such interpretations of the perseverance (or even reappearance) of populism do not do justice to some of its complexities. Another interpretation sees the new neo-populism as much more intimately connected to the effects of the recent history of neo-liberal restructuring in Latin America.

Roberts (1995: 83), for example, discusses the ways in which the application of neo-liberal policies in the context of redemocratisation in the region has produced a crisis in confidence within the traditional institutionalised forms of political representation. This breakdown in the traditional political system enabled the rise of neo-populist political figures who share (or rather exploit) the electorate's distrust of the political classes who, he suggests, 'are most likely to deposit their confidence in powerful men of action, in national "saviours" who promise to sweep away the detritus of the past and usher in a new social order' (ibid.: 83). Such considerations also serve to illustrate what Roberts sees as a much more complementary relationship between populism and neo-liberalism which 'is grounded in their reciprocal tendency to exploit – and exacerbate – the deinstitutionalisation of political representation'. Ultimately, 'the two phenomena are mutually reinforcing' (ibid.: 113).

This position is supported by Blokland (1996) who suggests three important affinities between neo-liberalism and populism:

> First, there are some similarities in their sources of societal support. They both appeal to unorganized, largely poor people in the informal sector, have an adversarial relation to many organized groups in civil society, and attack the established 'political class' as their main enemy. Second, their strategies of applying power coincide in relying on a strongly top-down approach and in strengthening the apex of the state in order to effect profound economic reform and to boost the position of the personal leader, respectively. Third, neoliberalism and neopopulism have some similarities in their distribution of costs and benefits. Neoliberal adjustment imposes particularly high costs on organized groups in civil society, which neopopulist leaders seek to weaken. In contrast, by ending hyperinflation and enacting targeted antipoverty programs, it benefits poorer sectors, which neopopulists court.
>
> (Blokland 1996: 10)

Such tendencies are also apparent within the use of the more recent targeted social programmes where populistic leaders have been able to simultaneously blame economic austerity upon external circumstances, imperialism and the errors of previous discredited political leaders, and enjoy the political benefits of targeting social measures towards client groups. These issues have profound implications for those seeking to oppose neo-liberalism and generate support for alternatives in that 'it warns against the comforting assumption that neoliberalism is incapable of generating a broad base of political support, and thus will inevitably produce a popular backlash in favour of progressive alternatives. Indeed, it provides compelling evidence that neoliberalism is both a consequence and a cause of the weakening and fragmentation of the popular collective actors who are essential to any progressive alternative' (Roberts 1995: 115).

As the ideological complexion of regimes changes through the process of adjustment the state is restructured in other, less obvious, ways. Leftwich (1995, 1996) rightly asserts it matters less what formal political position the government in power takes or how it came to power so much as the nature of the state. Hence, Gills and Rocamora (1992: 505) are able to observe that 'The paradox of low intensity democracy is that a civilianised conservative regime can pursue painful and even repressive social and economic policies with more impunity and with less popular resistance than can an openly authoritarian regime' (505). However, there are instances such as Ghana where the adjustment programme was carried out by a military government through a closure of democratic channels.

In general during adjustment the presidential and executive branches of the state take over much of the decision-making which is bolstered by the repressive power of the military. In such cases conservative-technocratic politicians become leaders with the business class and the middle classes providing political support. For example, 'In Senegal, the authoritarian features of one-party dominance were reasserted in the face of political opposition' (Beckman 1992: 94) while in Côte d'Ivoire President Houphouet Boigny clamped down so heavily on opposition parties that a situation arose of 'multi-partyism without opposition' (Aribisala 1994: 140). The other assumption is that the military becomes the handmaiden of authoritarianism but in many cases it remains central, if not pre-eminent, to political decision-making. The crisis in Nigeria over the past decade is a case in point (Ihonvbere and Vaughan 1995). The situation in Latin America is little different where 'Successful adjustment is easier when there is a strong, centralized presidency able to overrule Congress and when the government controls the trade union movement and other sources of potential opposition' (Green 1995: 164).

Beyond this we need to analyse the ways in which state officials formulate and implement adjustment-related policies. In many Third World countries there has been a problem of institutional capacity in terms of implementing development initiatives (Hyden 1983; Rothchild and Foley 1983; Conyers and Hills 1984). These tendencies were clear where the state assumed the burden for development planning and implementation. Under neo-liberalism this has taken a different direction. Hutchful's analysis of Ghana is useful in this respect. In it he writes that 'What has emerged in Accra is a parallel government controlled if not created by the lender agencies' (1989: 122). This small, technocratic clique are generally placed in the finance ministries and 'formulate' policy in collaboration with Bank and Fund officials either resident in the country or who visit on missions. Loxley (1992) even cites a case where the World Bank drafted a response to one of their own directives. Evidence from other countries supports this notion of a parallel government which contradicts the calls for accountability and democracy in the liberalisation process (Callaghy 1990; Van der Geest and Kottering 1994; Aribisala 1994; Leftwich 1996).

Hibou (1999) extends this by arguing that 'The public administration finds itself marginalized since, on the one hand, the government has to deal with demands from its foreign partners which are increasingly radical and detailed on matters of financial administration and economic arbitration, while, on the other hand, it must concentrate its attention on the *small decision-making cells*, leaving the rest of the civil service to continue its path of ineffectiveness' (97, emphasis added). For her, SAPs exacerbate the civil service's tendency towards non-accountability and the illegal 'privatisation' of state services as gatekeepers increasingly act in self-seeking ways. So, despite the many good governance initiatives, the strengthening of technocratic 'cells' feeds the criminalisation of the state rather than increasing its probity.

The corollary of this increased centralisation and non-accountability is the problem of implementation. Institutional weaknesses have exacerbated the effects of adjustment policies which represents another paradox of the techno-cratic model. The logic of using central ministries for implementation is justified on the grounds that sub-national levels of the state are too weak to implement policies. However, the state institutions through which deregulation is taking place are also part of the political apparatus which will lose power to markets and therefore fight to protect their position. Some institutional 'weaknesses' are therefore more like filibusterism whereas others are genuine weaknesses because of poorly trained staff, sketchy data and inadequate technological infrastructure. For example, in 1984 in Niger responsibility for the privatisation programme was placed in two ministries leading to 'unhealthy and unnecessary rivalry between the State, Board of Directors and the Director-Generals of the compa-nies' (Akinterinwa 1994: 159). In Zambia confusion over the removal of maize subsidies was the result of the 'extreme administrative weaknesses of the Zambian state' (Callaghy 1990: 296).

Despite this centralisation of control we saw that the IFIs' vision of the state includes a measure of decentralisation and participation in the building up of civil society. I have challenged the broader assumptions regarding the ideolog-ical motives behind this. This section examines briefly the issue of local government, service provision and welfare. According to a World Bank-sponsored conference the role of local government is to 'Provide cost-effective municipal services through public, private or informal sectors [and] find ways to incorporate, where appropriate, the initiatives of the private sector, the informal sector and NGOs' (1990: 3). These initiatives centre on the introduction of user charges and involve little real understanding or dialogue with the users (Wunsch 1991).

Studies show that local government has been manipulated by central government and underfunded so that local communities tend to 'disengage' from this arena of politics (Olowu and Smoke 1992; Mohan 1996). Political issues arising out of the ensuing social hardship are not considered since it is the unquestioned application of market mechanisms which is important. Paradoxically, the revenue imperative leads to 'fiscal decentralisation' which penalises economically backward local authorities and thereby exacerbates uneven development. In this sense strengthening of local government and increasing fiscal accountability is used as a means of deflecting attention from the fact that these debilitating policy measures were devised and implemented centrally and undemocratically (Slater 1993).

A further major paradox emerges in the contradiction between austerity and legitimacy. As we saw in chapter 4, demand management policies tend to hurt the urban formal sector harder than the export-producing rural sector as it reverses urban–rural terms of trade. They also hurt articulate professional groups who work through state or parastatal organisation such as lawyers, lecturers and students. In reducing state expenditure the level of investment drops which can increase absolute and relative scarcity which further undermines policy intervention and it is the already fragile central state which is faced with the legitimacy backlash. In Latin America 'social and political unrest have become the norm, as continued opposition to the impact of structural adjustment has sputtered and occasionally ignited in sporadic riots, strikes, rural uprisings, land takeovers' (Green 1995: 195). The consequence of lack of economic growth coupled with austerity measures is very often social protest against the government so that the IMF's illegitimacy is transferred to a lower political level which effectively insulates it from the contradictions it has unleashed. Hence, there is a blurring of national and international boundaries between the initiators and implementors of austerity packages. The most notable example of protest in Africa came in Zambia in 1986. Following an adjustment programme in 1985, maize prices increased and subsidies were partially removed, but the details were unclear, which resulted in severe shortages in the Copperbelt. The riots which followed saw fifteen people killed and protest spread to Lusaka where Kaunda reversed the subsidy decision and promptly nationalised the large flour mills (Callaghy 1990). Other studies highlight this tendency, especially among organised labour (Rothchild 1993; Oluyemi-Kusa 1994). As Barya (1993: 16) notes: 'SAPs are necessarily contradictory to the development and sustenance of democratic government.' Once austerity measures have been enforced they release the 'future crisis of "ungovernability"' (Gills and Rocamora 1992: 507).

The neo-liberal agenda is, by its very nature, highly centralised because it seeks to free markets which are viewed as ubiquitous and spaceless, unlike state apparatuses which are place-bound and spatially uneven. In this way deregulation of markets entails the re-regulation of political space and this leans towards authoritarianism. This then precludes any real redistribution of social welfare which increases social and regional polarisation in countries whose space economy is already skewed towards relatively immobile primary products. Another important point is that contrary to the zero-sum 'state or market' model some parts of the state are strengthened while others are trimmed. It is not simply a removal of all state functions. Those agencies involved in regulation tend to grow at the expense of welfare-oriented functions which become either privatised or given over to NGOs. Finally, the idea that the 'local' can become a site for empowerment ignores the ways that the state is able to manipulate and control local politics which I shall discuss below.

The IFIs respond: the issue of governance

In response to some of these political paradoxes there has been a recent emphasis on an amorphous concept of governance. The governance concept involves both democratisation and administrative reform. Like SAPs in general, this conditionality also raises important questions around sovereignty and autonomy as states are being 'forced' to be democratic. The IFIs however have very fixed notions about the nature of sovereignty, democracy, the state and civil society.

The reasons for the emergence of these ideas in the late 1980s were both theoretical and practical and set the context for the wider analysis that follows. First, in implementing the harsh conditionalities of SAPs, lenders were aware that there were vested interests within the state that were opposed to adjustment. These groups could and did upset the smooth implementation of adjustment programmes (Williams and Young 1994). In order to weaken these coalitions the lenders needed to bypass the state so their version of democracy with its stress on civil society was one means of 'legitimately' doing so (Leftwich 1996). Second, events in Eastern Europe not only gave weight to the argument that market distribution was more viable than state allocation but they removed the strategic-ideological counter-balance to excessive intervention (Hawthorn and Seabright 1996). Third, within the intellectual community there were a number of changes which enabled and legitimated these ideas. In the USA and Britain the Reaganomic and Thatcherite experiments were beginning to unfold, which set the standard for policy intervention. There was also a realisation that politics was important insofar as it could provide the checks and balances on the economy, especially in cases where excessive corruption was deemed to have hindered economic growth (Leftwich 1994). These factors influenced a new cohort of staff within the World Bank who were also working within a recently redesigned organisational structure (Williams and Young 1994). Fourth, there was the influence of pro-democracy movements in the countries concerned. Previously the strategic concerns of the lenders had been anti-communist consolidation and/or ensuring the profitability of foreign capital. Now these calls for democracy chimed with the other factors and gave added legitimacy to the lenders' agenda. I shall return to this below.

These factors conspired to effect a major change in political thinking. The publication of *Sub-Saharan Africa: From Crisis to Sustainable Growth* (World Bank 1989) marked a watershed in thinking about governance although the signs had been there from the early 1980s (World Bank 1983). At the most general level there has been a shift from the ideal that 'political development is dependent upon a combination of economic, social, and cultural "requisites" that are unlikely to exist in countries with underdeveloped economies' (Sklar 1996: 29). In the work of the political development school of the 1960s democracy could only follow economic development. However, the World Bank argued in the 1980s that 'political legitimacy and consensus are a precondition for sustainable development' (World Bank 1989: 60; World Bank 1991). As Leftwich points out, 'democracy is a necessary *prior or parallel* condition of development, not an outcome of it' (1993: 605, original emphasis). It is this key change that characterises the lenders' political thinking in the 1990s.

For the lenders governance relates to democratisation on the one hand and administrative reform on the other (Leftwich 1993, 1994; Baylies 1995). As Williams and Young (1994: 87) note: 'The Bank's promotion of civil society is linked to its promotion of accountability, legitimacy, transparency and participation as it is these factors which empower civil society and reduce the power of the state.' Democratisation is promoted as a way of weakening the excesses of the 'Leviathan' state because it transfers power to civil society. However, democratisation is reduced to an electoral process and as we have seen the state is much more than individual governments so you cannot easily remove unaccountable officials. On a broader note the conditionality of the governance agenda suggests that democracy is being imposed which opens up the whole issue of the accountability of the lenders. As Lancaster (1993: 12) notes, 'the Bank cannot openly admit the degree of its involvement in political issues without appearing to contravene the word and spirit of its articles of agreement' which state that 'the Bank and its officers shall not interfere in the political affairs of any member'. Barya (1993: 17) is even more condemning when he asserts that 'the very idea that people can be forced to be democratic and/or free is quite startling'. This apparent erasure of politics conceals the highly ideological role that the IFIs play.

The governance agenda also relates to administrative reform (Dia 1993). This is based upon a Weberian reading of the rational bureaucracy which is transparent and rule-governed. In concrete terms this involves cutting back on state employment, removing anti-reformers and streamlining ministries. In lender discourse the state is not envisaged as holding a significant stake in the means of production as distributive justice is given over to the market (Gills and Rocamora 1992). Additionally, state officials are to be held accountable to civil society rather than political elites (Dia 1993). Finally, the pressing need to stem the balance-of-payments problem and begin debt repayment means that revenue generation and cutting expenditure are paramount. Again, all these attack the state in significant ways and move it towards being a minimalist-technocratic state. However, as we shall see in 'deregulating', the state strengthens certain branches and re-regulates on the basis of a centralisation of power. The policies which flow from these ideas centre on privatisation of state-owned enterprises (Bienen and Waterbury 1989), the introduction of user charges for state services, and a variety of civil service reforms concerning retrenchment, pay scales and other incentives (Dia 1993).

The administrative use of governance relates to Lynda Chalker's claim that 'Above all, government must provide stable, predictable conditions in which to do business' (1993: 24); what the World Bank refers euphemistically to as the 'enabling environment' (World Bank 1992b). The sixth World Development Report made explicit the links between market reforms and administration whereby 'competitive markets permit the necessary flexibility and responsiveness and, because they decentralize the task of handling information, also economize on scarce administrative resources' (World Bank 1983: 53). Therefore, the scope of the public sector is narrowed or where it is maintained it should exploit the benefits of the market economy.

Brett (1988), Toye (1995) and Sklar (1996) note some of these moves should be welcomed as they can correct 'bad governance'. However, by divorcing these considerations from a wider analysis of political economy a number of important considerations are overlooked which have significant political and social implications. More generally the idea of governance is flawed because of its equation of democracy with Western liberal democracies, the detachment of democracy from social reform, the belief that states operate with a utopian Westphalian autonomy, and that the state should enable the economy so that administration is non-political. These in turn relate back to an artificial separation of politics from economics and a division of politics into three separate spatial spheres – the global, national and local.

One of the key elements in the governance agenda which ties these two aspects together is decentralisation and participation. This too has been approached from a rational choice perspective. Rondinelli *et al.* (1989) aimed to develop a 'new' political economy framework for analysing decentralisation programmes. The authors add privatisation as the main means of decentralisation which marked a shift from their earlier work which ranked deconcentration and devolution as the prime methods. The state is viewed in terms of either 'constraining' or 'enabling' and 'society' is reduced to the characteristics of people as consumers. By reducing people to consumers, they quietly ignore the other dimensions of decentralisation, namely the notion of participation in state decision-making, although the implication is that 'participation' means market transactions.

The sixth World Development Report makes the links between market reforms and administration even more explicit whereby 'competitive markets permit the necessary flexibility and responsiveness and, because they decentralize the task of handling information, also economize on scarce administrative resources' (World Bank 1983: 53). In this light decentralisation 'should be seen as part of a broader market-surrogate strategy' (ibid.: 123). In a related way when the major lenders talk of 'the local' and 'participation' they are primarily concerned with marketisation. For example, a recent World Development Report states: 'The *participation* of *local* businesses can also play a crucial role in decentralization, shaping incentives at the *local* level. ... Much of this began in *local* environments. Members of the business community often participated in *local* legislatures' (World Bank 1997: 123, emphasis added). Hence, those accounts which fail to explicitly acknowledge the capitalist basis for much economic activity and re-encode it as 'self-help' may inadvertently be opening up a space for capitalist social relations to take hold.

In reality decentralisation is highly contested for various reasons, not least of which is the power that central administrators fear that they will lose if lower tiers of government are empowered. In Ghana, for example, the central ministries were unwilling to devolve control to the districts for planning so that inefficiencies could not be overcome. In the Ministry of Education, the only moves in this direction concerned data collection for planning while the next phase of training District Education Officers never occurred. To date, the World Bank have withheld further funding to the education service until 'wastage' is reduced. Similarly, the central Ministry of Health has been restructured although this does not seem to have brought about the benefits at the sub-national levels. The basis of the reorganisation was downsizing and to produce greater efficiency within the ministry's current remit. The restructuring did not adequately plan around what is actually required to achieve decentralisation of health delivery. Indeed, there is a degree of scepticism about whether any of the central ministries are really prepared to decentralise authority. While the Ministry of Health has an inappropriate organisational blueprint for decentralisation, central officers will be willing to work towards it since its realisation will not threaten their authority.

A final point that we will flag up here, but return to in more detail in chapter 9, is that the civil society and the NGO fever associated with it are also fraught with problems. Hibou (1999), in talking of Africa, argues that the massive growth of NGOs has enabled unscrupulous actors to establish organisations which largely siphon off financial resources from donors and this represents 'the privatization of funds for aid and development' (ibid.: 99). She concludes that 'The promotion of NGOs leads to an erosion of official administrative and institutional capacity, a reinforcement of the power of elites, particularly at the local level' (ibid.). This suggests that the uncritical promotion of decentralisation, participation and civil society can be counter-productive and enable established powerholders to maintain or strengthen their positions.

Conclusion: resistance and new political spaces

Recently, attention has been paid to the ways in which globalisation and neo-liberalism have been or could be resisted (Gills 1997). The 'inevitabilism' of these discourses is rejected by seeking 'to produce, or at least make a contribution to producing, concrete strategies of resistance' (ibid.: 11). As globalisation is multiple so too must be these resistances. Chin and Mittelman (1997) argue that resistances will take three forms along the lines suggested by key thinkers Gramsci, Polanyi and Scott. They argue that 'Undeclared forms of resistance conducted individually and collectively in submerged networks parallel openly declared forms of resistance embodied in wars of movement and position, and counter-movements' (Chin and Mittelman 1997: 34). However, resistances are not always progressive as can be seen in the rise of extreme right-wing movements across Europe and the USA. Additionally, much resistance is largely reactive and should not be confused with a critical political conscience. The key then is to use resistances as a springboard into imagining and creating alternative futures (see chapter 8).

One line has been to stress resistance from 'below' (Falk 1997) which has seen a retreat into localisms underpinned by a philosophy of anti-development (Pieterse 1997). However, the danger is that the locality is reified as an hermetic social and political site which is at odds with the increasingly globalising tendencies of many economic and social processes. As Meiksins Wood (1995) comments with respect to debates around civil society: 'Its effect is to conceptualize away the problem of capitalism, by disaggregating society into fragments, with no overarching power structure, no totalizing unity, no systemic coercions – in other words, no capitalist system, with its expansionary drive and its capacity to penetrate every aspect of social life' (245). Pieterse criticises these strategies for simply seeking 'enclaves that provide shelter from the storm' (1997: 81), which then precludes the possibility of linking them together in a concerted global strategy.

For example, Marshall (1991) studied Pentecostal churches in Nigeria. These churches stressed the modernity of their views as opposed to the 'tradition' and 'heathenism' of Nigerian belief. They became popular from the early 1980s which coincided with the end of Naira Boom and era of adjustment, the so-called 'SAP-ped generation'. These religious communities created a number of responses to the difficult social and economic conditions. There was a certain egalitarianism in the congregations which broke down some of the sexism which had existed. They also stress 'conservative' sexual practices and frown upon infidelity which gives women more control in their relationships with men. The church protected its community and set up welfare programmes for born-agains. Paradoxically the members rejected the grotesque materialism which had accompanied the oil boom yet new business networks emerged between 'trustworthy' born-agains. The churches also provided new sources of capital from international Christian organisations which helped to form a new class of preachers. Critically, the churches became a protected space for organising anti-government activity although this had complex links with religious politics and the breaking of the Muslim stranglehold on Nigerian politics. While important political sites, these churches tended to promote communitarian approaches where local groups look after their own and look inwards rather than creating strong linkages with other civil society organisations.

So local resistance by itself cannot challenge global and, in this case, national forces. Hence, resistance must be 'localised, regionalised and globalised at the same time' (Chin and Mittelman 1997: 35). The linkages between scale and politics have become more complex, but more crucial, in these global times. We examine these alternatives in chapters 8 and 9.

6 The environmental aspects of adjustment

Giles Mohan

far from being a major source of environmental degradation in developing countries, adjustment policies appear, on balance, to have a bias in favour of the environment.

(Sebastian and Alicbusan 1989: 28)

Although Bank staff have been required to incorporate the general principles embodied in the concept of sustainability in preparing projects since 1984, it has not done so.

(Reed 1993: 173)

the Bank's strategy for structural adjustment lending has encouraged the export-led growth which has contributed to the environmental difficulties of African countries.

(Bush 1997: 505)

Introduction

Previous chapters have shown that SAPs are bound up in complex political, economic and social processes. In particular the proponents of SAPs were, as we have seen in chapters 1 and 2, so tied to an orthodox neo-classical package that they tended to ignore the wider ramifications of their programmes. The result was that as the messy realities came home to roost, new programme elements were 'bolted on' in an attempt to promote efficiency or to smooth and cushion the harsh effects of SAPs. For example, we have seen this in the areas of social welfare (chapter 4) and 'good governance' (chapter 5). Perhaps one of the most serious, but under-emphasised, instances of this was in the area of environmental impacts. It was only in the early–mid-1990s, a decade after SAPs were first implemented, that such considerations became critical issues.

It is difficult to determine the causal processes whereby the lenders began to acknowledge the environmental costs of adjustment, but some combination of pressures from local, national and international agents helps us explain why. Such pressures were largely routed through the popular and much-abused concept of 'sustainable development'. The political and intellectual climate was amenable to social forces which were agitating for an awareness that the resource base and the livelihoods it supported were under threat. The problem then became how to assess and separate the impact of SAPs on the environment from those effects generated by other policies. This chapter addresses both these issues. The first section traces briefly the genealogy of sustainable development and the linkages between environment and development. From there I unpack the relationships between SAPs and environment through both a theoretical discussion and case studies. In the following sections I look at the political responses to the gradual realisation that SAPs exacerbated environmental degradation. Such responses came from within the lender institutions as well as from social movements diametrically opposed to the adjustment agenda. Clearly, such political responses link us into the discussions in chapters 5 and 8.

(Un)sustainable development, the ideology of science and SAPs

This section examines the relationships between development and the environment. I argue that the lenders failed to address environmental issues because their epistemologies were neo-classical and based upon a positivist analysis of society and nature. The outcome of such ideologies was to neglect environmental resources or to treat them as relatively unproblematic factors in production.

Malthus, modernisation and the legacy of 'normal science'

A century and a half after Malthus's infamous and influential commentary on resources and society, a wave of independence spread across the colonial world. With this optimism came a renewed concern with the science of progress and faith in the power of development (Escobar 1995; Crush 1995; Leys 1996; Cowen and Shenton 1996). Former colonial powers, the super-powers and newly established post-colonial leaders all believed that underdevelopment could be ended through a process of modernisation. Orthodox modernisation posited a stages model tending towards equilibrium whereby increased agricultural productivity facilitated urbanisation and industrialisation. In this way development is linear with complementary industrial sectors, the so-called dual economy. Such theories were heavily Weberian in stressing the importance, albeit in a prescriptive and functional way, of cultural values. Hence, failure to develop arose from either poor entrepreneurial attitudes amongst the peasantry and/or inappropriate social institutions (e.g. communal tenure, weak social hierarchies). Such models therefore, like Malthus before them, blamed the victims for their poverty (Copans 1983).

In the 1960s population growth rates in these developing countries were high which prompted a number of studies in the early 1970s to talk of population 'explosions' and 'time-bombs'. The most famous of these studies are Ehrlich's *The Population Bomb* (1968) and *The Limits to Growth* (1972) model by Meadows *et al.* (known as The Club of Rome) (Corbridge 1986). Ehrlich, an ecologist, predicted that hundreds of millions of people would die of starvation. In a relatively unreconstructed Malthusian analysis he concluded that increasing aid to developing countries amounted to 'short-sighted programmes of death control' (quoted in Harrison 1993: 16). The only solution, as he saw it, was birth control which should be enforced if necessary. Hence, the onus for poverty was placed, once again, upon the poor and the global powers (and the dominant class forces that supported them) remained immune from criticism or blame.

Marxism and the Marxist critique of science

Marx and the Marxists who followed him rejected Malthus's ideas completely and viewed the population/poverty relationship as part of a wider analysis of the capitalist system. For Marx the poverty of the working class was a result of the capitalist system and could only be relieved by an over-turning of the system to a socialist one. However, despite this radical analysis Marx tended to stress labour and the wage system over other factors of production – in particular land and capital in the form of resources. It was his colleague Engels (1972, 1959) who mentioned the possibility of environmental degradation affecting capital accumulation (Redclift 1984). Subsequent Marxists such as Lenin and Luxemburg (and later the dependencia school) all emphasised the expansionary nature of capitalism via imperialism. This involved commodification of land, labour, and produce leading to an international division of labour mediated by mercantilism. This emphasis on global relations added an important dimension to ideas about the subjugation and alteration of 'natural economies', but it still ignored environmental considerations.

The neo-Marxist scholars added their critiques to the neo-Malthusian debates of the early 1970s. David Harvey (1974) attacked Malthus's positivistic scientific method and his biased concept of knowledge. Instead Harvey argued that we can:

1   change the ends we have in mind and later the social organisation of scarcity;
2   change our technical and cultural appraisals of nature;
3   change our views concerning the things to which we are accustomed;
4   seek to alter our numbers.

He believed that only in a communist society could all four factors be successfully changed and he criticised the 'natalists' such as Ehrlich who only focused on birth control as a means of resolving the resource/population dilemma. Keith Buchanan (1973) took a more internationalist line and argued that the notion of 'population crisis' shifted the blame onto the Third World when in fact the problem was with 'the white north'. His main criticism concerned the levels of consumption and hence the use and abuse of resources.

Beyond Malthus and Marx: the allure of sustainability

Throughout the 1970s ecological awareness grew, but it was not articulated into a coherent set of theories, policies or actions. By the early 1980s such holistic critiques began to emerge which supposedly left the Malthusian/Marxist dichotomy behind in favour of less ideologically charged programmes. The catch-all phrase became 'sustainable development' which was used so indiscriminately that it soon became an article of faith. Even a decade onwards, one commentator argued that 'It is far from clear whether sustainable development offers a new paradigm, or simply a green-wash over business-as-usual. In practice most commentators use it loosely and in an untheorised way' (Adams 1993: 207). For some it is defined as ecological sustainability while others define it as sustained growth in the economy. A third strand added that there are social conditions which influence the ecological sustainability or unsustainability of the people–nature interaction (Lele 1991).

In terms of the evolution of sustainable development there have been a number of landmarks which began in Stockholm in 1972 when the United Nations Conference on the Human Environment formally recognised the kinds of problems and concerns being discussed by the neo-Malthusians and their neo-Marxist detractors (Adams 1993; Reed 1996). The World Conservation Strategy devised at the beginning of the 1980s drew together themes from the previous decade which concerned the maintenance of essential ecological processes, the preservation of genetic diversity and, finally, the sustainable utilisation of resources. The strategy only addressed ecological sustainability and not sustainable development as such. The World Commission on Environment and Development published the influential *Our Common Future* (known as The Brundtland Report after its main author) which, like the Brandt Commission four years before, was loosely fashioned around the concept of Keynesianism. That is, that the regulation of the world economy could be mutually beneficial for all. In doing this it united growth in the global economy with environmental concerns.

However, the objectives of the report were vague in terms of practical action so that Lele (1991) was moved to say 'Where the SD movement has faltered is in its inability to develop a set of concepts, criteria and policies that are coherent or consistent – both externally (with physical and social reality) and internally (with each other)' (613). Many of the policies tend to be technical fixes which do nothing to address the broader context. For example, mainstream sustainability promotes free trade of commodities as being beneficial to producers, traders and buyers. In reality, trade is unequal, it encourages specialisation, leading to reduced biodiversity, and large producers benefit at the cost of smaller ones.

The 1992 Earth Summit in Rio marked the culmination of this political process whereby 'the international community formally embraced the concept of sustainable development as the standard for measuring development objectives and performance in both the North and South' (Reed 1996: 31). However, as with the Brundtland Report, the growth assumptions were not challenged and despite some organisational innovations the basic structures of power between North and South were left in place. Additionally, these emphases also marginalised social issues even further as faith was placed in trade with some level of international management.

In general, orthodox sustainable development does not challenge the economic growth of 'the North'. In keeping with Malthusianism, in most prescriptions, it is the sufferers who are called upon to change rather than the main perpetrators so that very little mention is made of reducing consumption in the North and effecting some form of redistribution. Second, the simplistic analysis of environment and economy does not account for the complexity of reality (Redclift and Benton 1994). The linkages between economics, politics, culture, technology and the environment are complex and situationally specific and require different models.

The political ecology of SAPs

So what are the implications of this for the impact of SAPs on the environment and our analysis of them? The discussions in the previous section highlighted that in the early 1980s, when SAPs were first devised, environment/population linkages were still being analysed in relatively simplistic terms where both 'environment' and 'society' were treated as discrete realms or that interactions between them were relatively straightforward and unilinear (Mearns 1991). Even when early sustainable development thinking emerged it tended to be either eco-centric or compatible with the growth assumptions implicit in the neo-liberal solution. Such problems stemmed from the flawed scientific assumptions of neo-classical economics (chapter 2) combined with their Malthusian and Darwinian social morality. In this section I look at the environmental impacts of SAPs as a ramification of this flawed approach. First, I look at why the lenders did not consider the environmental impacts of SAPs. Second, I introduce a theoretical framework which helps us analyse the inter-relationships between development and the environment in a more complete way. I then examine how these linkages elucidate the causal processes operating in and around SAPs before looking at a number of case studies.

Orthodox neo-liberalism and the environment

The neo-classical model on which adjustment is based rests upon adjusting the prices of commodities so that the balance between tradable and non-tradable goods shifts in favour of the former which in turn corrects internal and external imbalances. Mearns (1991) refers to this as the 'normal model' which will have environmental impacts depending upon local conditions. Overall, this model is highly generalised and is predicated on the belief that SAPs remain 'pure' during implementation and that implementation is completed in full. Second, the model, being based on systems analysis, assumes that relationships are causal, linear and quantifiable. The ecological and social units of the model are assumed to be stable and bounded and it is assumed that the overall system tends towards equilibrium. As such it is reductionist and unable to handle a process such as structural adjustment which, as this book stresses throughout, is unpredictable and complex.

The practical effect of these assumptions was a failure to incorporate environmental factors into adjustment policy. As Reed (1993) observes: 'Architects and proponents of adjustment programs failed to consider the impact of profound economic reforms on the environment of adjusting countries' (41). This arose from a number of inter-related factors:

- the World Bank and other lenders did not consider the environment to be a priority area at the time;
- those borrowing from the major lenders were also more concerned with pressing fiscal issues and did not specifically request funding for environmental protection;
- as we have seen, sustainable development in its broadest sense was not high in the public consciousness so that environmental protection was not an obvious policy element;
- environmental protection would incur further state expenditure which was the antithesis of adjustment programmes.

So if the epistemological criteria upon which adjustment is based did and do not provide satisfactory predictive or explanatory power then we need an alternative framework for analysing these impacts.

Political ecology and SAPs

Critics of the adjustment/environment nexus point to the need for more complex analyses. Mearns (1991) calls for a pluralistic approach which embraces complexity and treats the 'normal model' as a peculiar and minority variant of a generally unstable world. Redclift (1995) argues for a more society-centred approach which acknowledges the role of human agency in effecting environmental changes and, in keeping with this book's analysis, for a strongly political focus which does not treat the environment and its management as a discrete realm of activity separate from a 'political' one. In this section I want to introduce briefly the ideas of political ecology as a framework for analysing and politicising the relationships between development and the environment which breaks from the liberalism of much sustainable development discourse. It needs stressing at this point that analysing the environmental impacts of SAPs is difficult given the counter-factual trap. That is, how to assess whether the impacts would have occurred *without* an adjustment programme. In essence, how are we to separate out the adjustment from the non-adjustment effects? Such questions demand that we analyse both the direct and the indirect effects of SAPs as well as the short- and long-term impacts. This requires a dynamic approach which is attuned to both space/time and structure/agency (Benton and Redclift 1994). Such requirements can be met through the framework of political ecology.

As I discussed above, sustainable development made its debut in the 1980s although its elaboration on a world stage lagged behind the application of many SAPs. Around the same time Marxist-inspired analyses of environmental degradation began to appear (Blaikie 1985). Theoretically somewhat incoherent, this body of work became known as political ecology and has increased in theoretical sophistication (Blaikie and Brookfield 1987; Atkinson 1991; Bryant 1992; Blaikie 1995; Peet and Watts 1996). According to its earliest exponents 'The phrase "political ecology" combines the concerns of ecology and a broadly defined political economy. Together this encompasses the constantly shifting dialectic between society and land-based resources and also within classes and groups within society itself' (Blaikie and Brookfield 1987: 17). As a framework it brought together the concerns of political economy, ecology, and sociology and has latterly moved towards interactionist and structurationist perspectives (Redclift and Benton 1994; Blaikie 1995; Peet and Watts 1996).

Political ecology rejects economic reductionism of the neo-classical variety because this largely ignores ecological factors, neglects other sources of environmental change (e.g. the state) and underplays the role of social actors, especially the peasantry. Bryant (1992; Bryant and Bailey 1997) argues that a political-ecological analysis should encompass contextual factors, issues over access and the political ramifications of environmental change. I will explore the first two since the third is covered in the next section on responses to environmental degradation. While this political-ecology framework is useful it still tends to compartmentalise different processes which are in reality 'inter-mingled' in complex ways. Having said that it does draw attention to key social and political dimensions that neo-classical approaches ignore.

So how does a broadly political-ecological analysis help us to understand the complex processes at work in and around SAPs? Bryant (1992) identifies three contextual factors: state policies, inter-state relations and global capitalism. Although state policies cannot be separated from commercial interests, the state does have some autonomy so that it cannot be reduced to a 'black box' translating the interests of TNCs into policy. Focusing on the state as an actor also breaks with the economism of much environmental appraisal. State policies may reinforce commercial interests, such as a road scheme, or constrain them as in stringent regulations. Inter-state activity is differentiated into peaceful and violent activities. So-called peaceful activities include aid programmes which can again facilitate commercial interests or deepen dependency. Violent conflict is also environmentally damaging, creating much food insecurity and long-term environmental problems. Clearly, analysing the effect of SAPs is an example of applying this analytical category.

The third contextual factor is rather vaguely labelled global capitalism whereby the role and actions of TNCs are established in relation to environmental impacts. Such assessments should be historical and include the effects on state policy, individual firm strategies, the analysis of food systems and bio-technology, the ways in which national capital interacts with foreign capital and the actual environmental impact of firms. Again many of these effects are indirect outcomes of liberalisation and impact upon livelihoods in different ways. As with all these effects the contribution of SAPs is difficult to assess since many firms have been operating in developing countries for many years with a very poor safety and environmental record. One only need think of Bhopal to appreciate this. However, under SAPs we would expect the scale of TNC activities to increase although as we will see in subsequent sections there is no direct correlation between enhanced TNC activity and environmental damage. Similarly, it was often state-owned enterprises which had disastrous environmental records so that privatisation might, in some cases, reduce the damage.

The second main element of the framework concerns conflict over access and is focused at the local level. It is important to analyse tenurial systems and social institutions, gender relations and the household, and the access constraints on other socially disadvantaged groups. Such analysis is necessarily agency-centred, case specific and historical, but it should also not be separated from the structural factors as the two 'spheres' interact in complex ways. As Peet and Watts (1996) note, political ecology should avoid its earlier tendency to focus exclusively on 'the poor' as victims/agents of degradation since this ignores the 'obvious power of capital as a material force in degradation' (7) and ends up almost blaming the victims in true Malthusian style. In terms of structural adjustment we see numerous effects stemming indirectly from changes in relative prices. For example, the removal of subsidies on fertilisers can result in lower usage which in turn sees increased disease levels in crops but may, from another perspective, be a longer-term environmental benefit.

The third element of the framework is the political ramifications of environmental degradation. Although this is covered in the next section it needs stressing that the earlier political ecology was not particularly political. Any analysis should address two key questions: (1) do the socially disadvantaged bear the highest burdens of environmental change? and (2) to what extent does unequal exposure to environmental change modify political processes? We will see how such issues have been played out following SAPs and how supposedly 'environmental' conflicts have much wider political implications for the state and civil society. So much so that recent analysts talk about 'liberation ecology' (Peet and Watts 1996).

The impacts

Given that the assumptions upon which SAPs were based were largely environmentally blind we can assume that many of the environmental impacts were unforeseen and unintentional. But rather than get into an ultimately inconclusive debate about intentionality or otherwise it is better, following Reed (1996), to use the political ecology framework to examine the *direct* and *indirect* environmental impacts of SAPs. These revolve around the effects of relative prices and alterations to social organisation as a result of adjustment. In examining these impacts I have disaggregated the effects by policy area, but since this falls into the same trap of seeing these as discrete and unconnected I have also included short case studies which demonstrate the ways in which the totality of policy changes impacted upon the environment.

Direct

Price reform, as we saw, aims to align national prices with world prices and to bring national output in line with aggregate expenditures.

Exchange rate policy

Correcting overvalued currencies is central to most adjustment programmes as it attempts to encourage exports and temporarily suppress imports. In some countries this had beneficial impacts, especially for larger farmers, as it raised their earnings which could be reinvested in more environmentally sustainable methods. However, the rise in the cost of imports raised the price of key agricultural inputs so that in order to maintain productivity some farmers were forced to extend production. This occurred in Tanzania and Cameroon where production shifted to marginal lands. In other cases, where the revenue from exports was cut due to devaluation, some producers abandoned well-managed plantations and began to grow food crops on marginal lands (Winpenny 1996). Although it is difficult to generalise it was larger farmers who could bear the relative changes in prices and could cover the costs of inputs from export revenue.

Trade liberalisation

The impact of trade liberalisation again depends upon the country concerned and the nature of the world markets in which it trades. One effect in Jamaica, Venezuela and, as we will see below, Ghana, was to increase the flow of foreign investment into the extractive sectors. For example, in Jamaica 'Bauxite/alumina production ... has caused dust pollution and the problem of red mud disposal. Insufficient land restoration affects the productivity of farm lands and the salinity of water in the area owing to the volume of water needed in production' (Markandya 1996: 195). Paradoxically, heavy polluters such as mining companies require strict regulation, but this coincided with the cutting of support to state agencies.

In the industrial sector trade reform encouraged export production in key sectors such as pulp, paper, cement and petrochemicals. These tend to be larger-scale firms with the worst pollution records. It was hoped that by encouraging small-scale firms, which are environmentally less damaging, aggregate degradation would decline. As we shall see with small-scale mining in Ghana this is not the case as 'small firms' lack of technical know-how, could, in aggregate, increase the environmental damage' (Reed 1996: 308). In agriculture the impacts are also mixed, with some countries, such as Cameroon and Vietnam, diversifying their range of exports while others have suffered from declining output and competition from imports. In Venezuela, the rise in beef prices encouraged an extension of rangelands which resulted in deforestation while subsidy removal and the rise in import prices prevented many farmers from investing in sound production techniques. The effect was to reduce the range of agricultural commodities for local consumption and increase food imports. This further squeezed small, subsistence farmers who were forced to exploit more marginal lands and/or deforest new land. Such a pattern is typical across adjusting countries where 'Commerical farmers are able to shift crops, absorb increases in input prices, and adjust to new marketing arrangements tied to external markets far more effectively than small producers' (Reed 1996: 310). We will see in the subsequent section how this differential ability to 'absorb' and 'adjust' has engendered political protest in rural Mexico.

Other key sectors affected by trade liberalisation are forestry, tourism and transport. The forestry sectors were either targeted directly as export sectors or were affected by the substitution for other products following rises in the costs of imports. In Cameroon, the forestry sector was targeted as a key area and a Tropical Forestry Action Plan was developed to ensure sustainable extraction. However, due to poor logistical support, weak institutional capacity and corruption, the plan is not being enforced, resulting in indiscriminate logging (Winpenny 1996). In other countries the rise in costs of imported fuels such as kerosene saw consumers substitute them for cheaper local alternatives such as fuelwood and charcoal.

For many developing countries tourism is seen as a major potential income-generating activity. Tourism has the potential to benefit the environment if income is redirected towards sustainable management but under SAPs the impacts have often been the opposite. In Zambia, for example, the increased poverty caused by austerity has led to an increase in poaching as a source of income. This illegal activity has reduced big game numbers dramatically while the exacerbated decline of the National Parks and Wildlife Service has made it harder to regulate and prosecute these poachers (Cromwell 1996).

Transport has been affected in contradictory ways. In many countries the removal of subsidies on fuel has led to price increases and suppressed consumption. On the other hand liberalisation allows for the freer import of new and used foreign vehicles into these countries. Many of the older vehicles are poorly maintained and highly polluting. In Mexico, the state subsidised petrol and other energy sources since the Second World War. This resulted in massive problems of urban pollution as factories substituted energy-intensive production for labour-intensive techniques and road transport was cheap. In Mexico City 85 per cent of airborne pollution originates from motor vehicles. Adjustment could decrease urban pollution, but under NAFTA American firms have been hopping the border to exploit cheaper labour and more lax environmental laws which negates any benefits or, indeed, worsens the situation (London Environment and Economics Centre 1993; Redclift 1995).

Fiscal policy

One of the central elements of any SAP is the reduction in public goods and services, or their repricing. In nearly every case these policies impacted worse upon the poor which, in turn, places additional pressure on a country's natural resource base. The reduction of subsidies has diverse impacts as we have already seen. In some cases raising the price for agricultural inputs could reduce the use of harmful pesticides but in the face of falling income, many poorer farmers extend cultivation onto ecologically fragile areas. Another important element in fiscal policy is the reduction of extension services as governments reduce their expenditures. These services are not simply technical but include small-scale credit and inputs. The net result has been declining productivity. Additionally, the monitoring and regulation functions of these services suffer so that abuses of natural resources, such as timber, go unchecked. We will see in the next section how in response to this loss of local monitoring capacity, decentralisation has been advocated as an alternative. However, such local participation is no guarantee of success as local people, already burdened by heavy work loads, are being called upon to take up additional duties.

Privatisation

The experience of privatisation varies across countries. In African countries one of the first targets for divestiture were the state-run marketing boards. Although the impact on prices was welcomed by many farmers, there was increased confusion in the short term over marketing systems. The effect of this uncertainty was to dissuade some farmers from growing cash crops which in turn lowered their income which necessitated working marginal land.

Indirect effects: social reorganisation

Social

One of the key issues is that SAPs generally increase poverty and social polarisation. As we saw in chapter 4 this, in turn, forces poorer groups into various 'coping strategies' which often means over-exploiting natural resources which are seen to be 'free and available'. Incidences of poverty are related to two key processes within adjustment – the effect on labour markets and the withdrawal of social services (Reed 1996). As we saw in chapter 2, the stabilisation phase of SAPs places downward pressure on wages, especially in the state sector. Even more severe are the wide-scale retrenchments which release large numbers of government workers into the labour market. With limited employment opportunities many of these workers are forced into the 'informal sector' or combine several jobs in order to secure an income. One effect is that in peri-urban areas families have started market gardening for sale in the cities. In rural areas some families have resorted to hunting and charcoal production. This has been exacerbated, in some African countries at least, by a general shift in migration from urban to rural areas. However, in Tanzania and Jamaica, harsh agricultural conditions for poorer farmers have forced them to move to cities in search of casual employment which then places added burdens upon urban services such as housing, water and sanitation. Allied to this is a reduction in social services or their repricing which differentially impacts upon the poor who lack the disposable income to pay for education, water, etc. In particular, women bear the brunt of this as farmers and maintainers of the household.

Institutional

While much of the emphasis of SAPs was on getting prices right internally and externally, there has also been much made of institutions. We saw in chapter 5 that structural adjustment involves substantial amounts of 'rethinking the state'. Questions of environmental regulation are vital given that liberalisation of economies also results in cutting state expenditure at a time when capacity building needs to be at its most intense (Reed 1993; Redclift 1995). Numerous studies have highlighted the administrative weakness of the state in developing countries so there is a paradox between environmental goals and 'implementational capacity'. In Jamaica the environmental budget was cut by 50 per cent which then filtered through to logistics, personnel and infrastructure.

One response as we will see below is to emphasise the role of civil society in the management of environmental resources. However, in some countries the harsh effects of SAPs have weakened community ties, which makes the likelihood of self-help solutions less likely. Equally, the decentralisation of service provision aims to shift the fiscal burden away from central ministries, but the capacity at this level is weak. In particular, many local governments are still burdened by colonial structures which tie them into line ministries. Given that environmental degradation is, as we have seen, complex and multi-causal, such sectoral rigidities work against holistic environmental planning and management.

Case studies

In this section I have detailed two case studies which exemplify the complex inter-relationships between these various processes.

Small-scale mining in Ghana

Ghana's structural adjustment efforts of the 1980s have been hailed as one of the more successful attempts in sub-Saharan Africa (Leechor 1996) with an increased focus on the natural resource sector, especially agriculture, forestry and mining. In the mining sector in Ghana a number of laws were passed promoting artisanal mining whereby these previously illegal operations were to come under state regulation. This led to the Small-Scale Gold Mining Law (PNDC Law 218) of 1989 and the establishment of a semi-private Precious

Minerals Marketing Corporation (PMMC) which buys the gold and diamonds and captures the revenue from this sector. The combined impact of encouraging small-scale and multinational mining companies has been a dramatic increase in output (Republic of Ghana 1993). In terms of small-scale mining 279 licences were granted between 1989 and 1991.

The environmental impacts of mining are varied and depend upon the ecological character of the mining site. Additionally, the mining process is not simply the extraction of the raw mineral, but a complex series of concentration and refinement. The small-scale mining known as *galamsey* has been the traditional form of mining in Ghana. Following the 1989 law, production by these companies has increasingly flowed into the PMMC and the sector employs 50,000 people (Nyamekye 1992). These operations are also significant for their role in showing the way for larger prospectors which produces local conflict and can therefore frustrate efforts to regulate the industry. Most *galamsey* activity takes place on or near the large concessions with the large companies employing security staff to police these illegal operations (NSR 1991).

An examination of these operations shows that they are under-capitalised and lacking technical expertise (Asante *et al.* 1993). Equally important is that ownership is separated from those doing the mining who operate under a tributary system (Ofei-Aboagye 1992; Dzakpazu 1992; Jumah 1993) where miners retain part of their find as payment from the mine-owner who keeps the rest. In general it was found that:

1　as exploitable gravels and soils are used there is a tendency to mine lode (rock) which incurs the use of explosives with the associated safety risks;
2　the companies lack finances and often abandon sites without rehabilitating or stabilising them;
3　the concessions are not properly prospected so that ore deposits are not clearly delimited which leads to excessive digging and unnecessary environmental damage;
4　safety standards and regulations are weak to non-existent.

More specifically this leads to the following environmental problems:

1　widening of river banks and loosening of soils;
2　removal of vegetation and overburden;
3　digging of deep trenches which damage rocks and cause subsidence;
4　plants of economic value are not collected;
5　contamination of rivers during beneficiation;
6　contamination of rivers by large work gangs and bank erosion.

Much of this mining is illegal and this has secondary effects. Not all of the *galamsey* mining takes place in remote areas. Newspaper articles show that it encroaches onto roads and residential areas. Indeed there were reports of people falling into open pits and dying. The lure of quick wealth has led to young men taking up mining instead of farming so that food production has decreased which has been exacerbated by an influx of miners from other areas of Ghana.

These operations have not been closely regulated for a number of reasons. First, the mines have been in place for hundreds of years and 'There has never been a mine developed in Ghana which has not first been discovered by a native surface working' (Minerals Commission 1991: 11). Given their role in prospecting, the larger mines do not wish to eradicate this tradition. Second, the concessions cover large areas and the mines are numerous and dispersed which makes them hard to monitor. Third, these operations are embedded in the local economy and entire communities benefit from their existence. Hence, monitoring and apprehension by local citizens are unlikely. Small-scale mining forms part of a huge semi-informal economy which is 'part and parcel of an entire system of production, exchange and distribution around which particular forms of social and political relations are embedded' (Chachage 1993: 89; Hollaway 1996). These 'integral production systems' are not 'alternatives' to the 'formal' economy because such dualisms (formal/informal) underestimate the complexity of African social formations in which a range of economic arrangements exist. Related to this is the fact that employment in SSM is large, though difficult to quantify precisely, which is a disincentive to regulate such activities and risk closing them.

Fires in SE Asia

In early October 1997 an airbus A-300 operated by Garuda Airlines crashed into a hillside thirty kilometres southwest of Medan, the north Sumatran capital. There were no survivors from the 234 passengers and crew. Although the causes are still being investigated, the most obvious reason was the poor visibility due to the smoke which had been shrouding the entire region for the past few months. The smoke came from numerous forest fires which were burning trees and, more problematically, peat deposits and surface coal measures. The worst affected areas were the islands of Borneo, Java, Sulawesi and Sumatra where 300,000 hectares of land had been affected by fires; satellite photos indicated that the Indonesian total was around 500,000 hectares. At this time the air quality was appalling with around 50,000 people being treated for respiratory disorders. The air quality in Kuching, the main city of the state of Sarawak, was 839 points where the absolute maximum is 500 points and 300 is akin to smoking twenty cigarettes a day.

The causes were hotly contested and pointed to the effects of rapid economic growth and bad governance. These fires had occurred for a long time, but they had been increasing in frequency and intensity over the past few years. The official explanation had always been that small, subsistence farmers were responsible due to their careless 'slash-and-burn' methods. However, Noerid Radam, a professor at the Lambung Mangkurat University, commented that these farmers are 'so expert in their techniques that it is inconceivable that their fires would spread beyond their small fields' (cited in Hajari 1997: 42). The other contributory factor was the alterations in weather patterns caused by

El Niño. The region had been suffering from a drought and the forests were extremely dry. However, this exacerbated an ecological crisis with roots in Indonesia's longer-term development strategy.

The Indonesian government had been pursuing an aggressive economic programme for a number of years. The strategy was based on attracting inward investment and exploiting the country's abundant natural resources such as timber, plantation crops, minerals and its natural beauty as a tourist asset. For example, since the late 1960s the infamous Freeport McMoRan gold mine in Irian Jaya has displaced 40,000 indigenous people and destroyed the local environment. The major lenders have supported the economic 'miracle' and the so-called tiger economies were, until recently, held up as models of modernisation and development. However, as Chip Barber of the World Resources Institute commented: 'The fires are the underbelly of the free-for-all economy that the World Bank, the IMF and the bankers have underwritten and held up as an example to other developing countries' (cited in Vidal 1997: *Guardian* 8 Nov.). These underlying processes, then, created the conditions in which a number of more immediate factors brought about ecological devastation.

The first major cause was the drive to increase oil palm production and the competition between Malaysia and Indonesia. Malaysia is the world's top oil palm producer, but Indonesia is attempting to usurp it. In an effort to protect their position and increase profits, the Malaysian companies are also establishing plantations on Indonesian soil as there is less of a land shortage there. In Indonesia the cost of production is lower than in Malaysia, $150 compared with $250 per tonne. It is these companies that have been burning forests to clear way for oil palm plantations. Perversely, the fires make this competition more intense as the fall in output will raise the world market value for the next few years.

Although the Indonesian government were aware of the fires their response was lukewarm. In mid-September they threatened to revoke the licences of 176 companies but at the ultimatum of 3 October they revoked only 29. The problem is the close and nepotistic ties between the major palm oil companies and the government. For example, President Suharto's confidante Sudomo Salim is one of the biggest operators which has 'shielded them from regulatory action' (Hajari 1997: 42). During the fires Abdurrahman Wahid, the leader of Indonesia's largest Muslim organisation, said: 'Local officials are doing nothing at all. ... The government and the army should start doing something quickly' (cited in Thoenes 1997: *Financial Times* 26 Sep.). So one of the key causes was the crisis of governance and the weak environmental legislation in the country, fuelled by inter-state competition and complex patterns of trans-border investment and migration.

The second key factor was the major developmental projects being undertaken in the region. President Suharto, the then President, had initiated a massive agricultural scheme in Borneo on the Kahayan river. The aim of 'The Project', as it is known locally, is to drain one million hectares of peat swamp and plant it with rice. This necessitates resettling 30,000 farmers from Bali and

Java to grow the rice. The first move is to drain the swamp and then to clear it using fire. As one local resident exclaimed: 'The fires are from the Project. ... They just let it burn, and then they started clearing the land for rice fields' (cited in Peel and Thoenes 1997). Similarly, in Malaysia, the Bakun Dam project aims to develop hydro-electric power, but even if the dam was unprofitable, the developers have promised sizeable logging contracts nearby.

The further effects of the fires are also serious. The combined effect of drought and the fires, which reduced sunlight at a crucial growing period for many crops, has been to reduce agricultural output in both staples and export crops. Rice, coffee and palm oil output are all down which in the short turn worsens the balance-of-payments position as export revenue falls. It also fuels inflation as food crops, rice in particular, have to be imported. Tourism in the region has also been hit which is serious for countries like Singapore where it constitutes 8.5 per cent of GDP. Additionally, a wider governance issue was exposed during the crisis. The Association of South East Asian Nations (ASEAN) which aims to co-ordinate development across the region was unable to react to the fires. Partly through politeness, and partly due to the mutual complicity of the individual leaders, little was done until international pressure was placed on them. This inability to act together exposes deeper rifts between them and questions the future of regional governance. Recently the Asia-Pacific Economic Co-operation (APEC) forum further liberalised the trade in forest products which could provide new markets for Indonesian exports and further exacerbate the problem.

Responses to environmental impacts

The preceding sections showed that the lenders did not consider the resource and environmental impacts of SAPs when they were first implemented. Although separating out the SAP-induced effects from other effects is difficult, there has been a realisation that environmental damage has increased and that regulatory mechanisms needed to be factored in. Such moves were in keeping with the lenders softening on the issue of state regulation which had been the demon of early SAP ideology (see chapter 5). So from the early 1990s moves have been made to incorporate environmental management solutions into programme elements. These, as we shall see, were in keeping with SAP thinking such as privatising land ownership and improving the capacity of state institutions to carry out environmental protection. However, more radical responses have emerged which link environmental politics to wider debates around citizenship, identity, democracy and sovereignty. This section considers both the institutional and societal responses and I conclude by raising questions about the likelihood of them changing the (ab)use of environmental resources and bringing about more complete and locally appropriate forms of sustainable development.

The lender's responses

This section leads into the discussion in chapter 7 regarding modifying the adjustment paradigm without fundamentally challenging its rationale. The 1992 World Development Report (World Bank 1992c) presents a somewhat optimistic analysis of the environment-development nexus. The report 'identifies the conditions under which policies for efficient income growth can complement those for environment protection [while] policies that are justified on economic grounds alone can deliver substantial environmental benefits' (1). From this opening the report suggests a number of 'win–win' scenarios although the supporting analysis is firmly Malthusian (Williams 1995; Bush 1997). For example, rural populations are blamed for 'poor stewardship due to poverty, ignorance, and corruption' (World Bank 1992c: 19) and 'When people have open access to forests, pastureland, or fishing grounds, they tend to overuse them' (ibid.: 12). Additionally, 'The slowly evolving intensification that occurred in the first half of this century was disrupted by the sharp acceleration of population growth in the past four decades [which] delayed the demographic transition and encouraged land degradation and deforestation' (ibid.: 8). At no point does the report acknowledge the longer-term effects of economies skewed towards production of raw materials for export or the impacts of multinational capital.

Similarly, the Bank does not specifically acknowledge the impact of its own SAPs on the environment although they hint that 'well-intentioned policies have been thwarted by other policies that pull in the opposite direction' (ibid.: 13). The only admission they make is in the highly visible area of infrastructure, such as dam building. Here the report acknowledges that development agencies often failed to 'take environmental considerations into account or to judge the magnitude of the impacts' (14). While valid, such honesty masks the other indirect effects of IFI policies while the solutions it recommends, as we shall see, relate to the relatively safe and fashionable idea of participation in project design.

The proposed solutions to these problems revolve around the relationships between market and state, but are broadly consistent with SAP logic. On the market side governments are encouraged to remove market distortions by eliminating subsidies and change behaviour through various incentives and restrictions, such as 'polluter pays' schemes (OECD 1993). Their answer to the management of resources is firmly rooted in the price mechanism and private ownership whereby land rights are clarified and user charges introduced. On the regulatory side the agencies are to have their aims clarified and capacities built up to ensure monitoring and enforcement are efficient and effective. Mearns (1991) describes these changes as 'reformist' and as unlikely to lead to significant improvements because, based as they are on environmental economics, they are premised upon neo-classical, positivist assumptions. As such they are simply 'bolted on' as supplements to the normal model.

International governance: the GEF

One of the key concerns around global environmental governance is that the various international organisations which exist, including the BWIs, should co-ordinate their activities more closely (Griesgraber and Gunter 1996a; Imber 1997). This is seen as the only way to handle the complex, inter-sectoral nature of environmental degradation and to address its increasingly international dimensions. In 1990, moves were made to establish the Global Environment Facility (GEF) which aimed to help developing countries protect the global, as opposed to local, environment. Although a major inter-agency initiative, the World Bank, UNDP and UNEP were the three implementing agencies, with the World Bank as the central trustee. Disbursements are made on a grant or concessional basis (World Bank 1992c; Jordan 1995). It is not surprising, given the essentially selfish motivations on the part of developed country govern-ments, that the GEF was commandeered by them (Reed 1993). As with most of these global governance initiatives (e.g. WTO, MAI), developing country representatives were only brought in to consultations once the basic working parameters had been set.

This self-serving agenda was borne out in the ways that developing country priorities around poverty were not addressed. Also, given that the GEF only funds the 'incremental costs' associated with protecting the global environment, it did not challenge existing interventions but simply 'piggy backs' on them and could 'merely be used as a cloak to "greenwash" pre-existing projects that are inherently flawed' (Jordan 1995: 308). Finally the GEF was placed in the hands of the World Bank, precisely because it has a weighted voting structure, unlike the UN which has a one-country-one-vote system which gives more power to the developing countries. As such the early phase of the GEF suffered from a more general malaise regarding the UN system with a 'head on collision between voting power and financial power. Voting power can control agendas, but financial power determines programme implementation and outcome' (Imber 1997: 225). Such undemocratic practices are at odds with the NEAP process which we examine below.

By 1992, when the Rio Summit was held, the GEF management made limited concessions to developing countries, but more importantly it was given the go-ahead as a key mechanism for co-ordinating global environmental gover-nance. Since then the GEF has moved tentatively towards democratising itself and this has raised the likelihood of the World Bank moving more in line with the rest of the UN system in terms of voting and decision-making. Additionally, despite being bound to the 'incremental costs' principle, the definition and means of calculating this are unclear. Given that the GEF still commands rela-tively insignificant funds, its greatest impact may be in pushing forward the idea that environmental degradation will not be ameliorated by single-agency solu-tions in democratically limited ways.

National institutional strengthening: the NEAP process and decentralisation

One of the major innovations in environmental policy has been National Environmental Action Plans (NEAPs). Although NEAPs are country-centred, locally generated and not directed by the World Bank, they have received strong support from major multilateral and bilateral donors (Falloux and Talbot 1993). The World Bank contributed funds and consultancy expertise in the formulation stages, but the authors of a major study on NEAPs are at pains to stress that 'so too did those from other donors' (ibid.: xv). Clearly, the political issues of sovereignty and ownership (see chapter 5) have played a major part in the rhetoric, at least, of these initiatives.

NEAPs have been pioneered in Africa and are seen as a long-term process rather than a short-term technical fix. Falloux and Talbot (1993) describe them as aiming 'to provide a framework for integrating environmental concerns into a country's economic and social development, and to embed that framework in the fabric of the government and peoples so that it is their process' (19). They hail the NEAP process as special because 'It has been created and developed in Africa [and seeks] authentic African solutions' (xiv–xv). NEAPs are characterised by being demand-driven, African in origin, an in-country process, involving broad national participation, a long-term process, an holistic approach, focusing on underlying causes, linked to action, having government commitment and involving donor participation. Although few studies have been conducted of NEAPs they appear promising in that they acknowledge the inherently political problems embedded within management solutions. The question of ownership is important if NEAPs are to avoid becoming simply another item on the list of conditionalities. For example, Falloux and Talbot (1993) described Ghana's NEAP as being the most free from foreign technical assistance which is a problem that has plagued Ghana and most adjusting countries (Hutchful 1989; Loxley 1990).

Another interesting aspect of the NEAPs and recent lender discourse is their emphasis on participatory institutions and decentralisation. The 1992 World Development Report (World Bank 1992c) states that 'Making choices between economic and social benefits and environmental costs often requires subjective judgements and detailed local knowledge' (15). This emphasis on participation has also entered into other policy areas, most notably poverty assessment (Norton and Stephens 1995). The NEAP process also aims to 'involve the general public throughout the country involved, often through consultations with district, ward and village representatives' (Falloux and Talbot 1993: 23). The assumption is that 'better' and more equitable solutions will be found to the problems of local environmental management. However, the discourse of local choice can be interpreted within the context of marketisation whereby private ownership of land is portrayed as the only rational and sustainable form of management. Similarly, the emphasis on the local state and the use of blanket terms like 'the local' can conceal wide social and economic disparities within communities. For example, in Kenya, Woodhouse (1997) found that the 'current policy emphasis on increasing the power of local authorities tends to obscure important and rapidly-evolving power relations between social groups' (545). There is, then, a need to deconstruct 'the local'. A related point is that such prescriptions can focus too much attention on supposedly local entitlement issues which then obscures the important effects of the central state. Fragmenting societies into multiple localisms may mean that progressive reform of the centre does not occur.

Popular movements and liberation ecology

Another area of political responses has been outside of the formal institutions of governance. We saw above that the political ecology framework incorporates political ramifications of environmental degradation which has seen a focus on social movements and civil society (see chapters 5 and 9). The emergence of a global environmental consciousness has been an important part in the growth of 'global civil society' (Lipschutz 1992; McGrew 1997) although such movements should be disaggregated because the agendas of the groups which comprise such transnational networks may not be the same. In the West the protesters are often not involved directly in livelihood struggles whereas those in developing countries usually are (Redclift 1987). Although this opens up debates about how we interpret 'values' and 'livelihoods', we must avoid projecting our own agendas onto others or interpreting their agendas in terms of our concerns.

Peet and Watts (1996) argue that much political ecology fails to have 'a serious treatment of politics' (10) and that we must examine politics in its broadest sense. They talk of the 'panoply of political forms – movements, domestic struggles over property rights, contestations within state bureaucracies – and the ways in which claims are made, negotiated, and contested' (ibid.: 11), which takes local and regional forms, rather than some universal conception of an environmental consciousness and a shared agenda. The same authors talk of 'liberation ecology' rather than political ecology, because they believe such political action can be emancipatory and transformative. As such they 'recognise the emancipatory potential of what we call the "environmental imaginary" and to begin to chart the ways in which natural as much as social agency can be harnessed to a sophisticated treatment of science, society, and environmental justice' (ibid.: 13). However, linking such political processes to an agenda which seeks to transform or overturn the adjustment paradigm will not be easy, as the following case study shows.

In the 1990s in southern Nigeria the Ogoni people mounted a resistance campaign against the Nigerian government and the oil corporation Shell. The roots of the problem are deep and cannot be simply attributed to adjustment. The Nigerian state is involved in, and mediates, the global oil production while Bretton Woods-sponsored adjustment deepened social unrest (Obi 1997). Although centred around environmental degradation, the resistance movement was both place and ethnically based because the Ogoni people were being perse-cuted by the Nigerian state and, allegedly, indirectly by Shell itself. Hence, much of the resistance was reactive. The Movement for the Survival of the Ogoni People (MOSOP) called for greater political recognition within the Nigerian Federation, recompense for the damage caused by years of oil production and broader processes of self-determination. Initially, MOSOP was successful in campaigning with international organisations such as Survival International. By the mid-1990s they were so vociferous that the government attempted to silence them through repression. This reached its height in 1995 when the government imprisoned the MOSOP leaders on trumped up charges of organ-ising local killings. In November these nine MOSOP leaders, including Ken Saro-Wiwa, were executed, which caused international outcries but only limited official sanctions. It seems that the opposition has waned since then. This example shows that local resistance by itself cannot challenge global and, in this case, national forces. There is therefore a need to 'break out of the local ... primary or exclusive emphasis on the local can also lead groups to become colloquial and blinkered to other acts of resistance around the world or even their own regions, leaving them exposed to defeat or even destruction by not building sufficient social alliances' (Amoore *et al.* 1997: 190).

Ways ahead

The conclusion is tentative and relatively short as we wish to address similar concerns in the final chapter. The first area where changes can be made is in the epistemological and theoretical basis of the environment–development nexus. One way to do this is to revalue local knowledges as alternative epistemologies (Redclift 1995), although Escobar (1996) warns that this can be used as a means of control. However, while positivist and reformist policies pervade the IFIs it is unlikely that complexity will be revealed. Second, theories must reject attempts to focus on economic growth as a means of poverty removal and/or environmental sustainability. While growth is an important requirement it does not necessarily alleviate poverty and so some type of reform or redistribution is needed. Related to this, then, is a realisation that poverty and environmental degradation have structural, technological and cultural causes so that 'sustain-ability' has multiple dimensions. Such a recognition demands greater sensitivity to space and time (Reed 1996) which would then allow us to explore what patterns and levels of resource demand and use would be compatible with different forms or levels of ecological and social sustainability (Lele 1991). For example, Reed (1996) urges us to differentiate between key types of economies and their role in an international division of labour.

While there is an obvious need to continually rethink sustainability, such issues should be embedded within the existing political economy. Economically there is a need for fair and proper resource pricing (e.g. energy taxes, the aboli-tion of transfer pricing) as well as just access to resources. For example, many people in LDCs (Less-Developed Countries) are alienated from the land and have little incentive to maintain it so that land reform is crucial. Rather than a continued commitment to growth for poverty alleviation there needs to be greater investment in human development (e.g. education, health services, nutrition) as well as stabilising populations, but only through sensitive and carefully administered family planning programmes. Environmentally, commitments must be maintained to improve technology development and co-operation, stabilise the climate (e.g. reduce greenhouse emissions and pollution), sustain agriculture (e.g. reduce chemical use, rural support and credit, etc.) and conserve biodiversity. Politically, in addition to these local and national concerns, are more internationally focused ones around democratic reform of the multilateral lender institutions. Much international diplomacy is still biased away from developing countries so that the UN system in particular must be reformed. A key issue around which this could occur is that of debt relief.

Part III

Alternatives to adjustment

Introduction

> Abraham Maslow once said that if the only tool that you have is a hammer, everything begins to look like a nail. The prescription of liberalizing all conceivable markets in countries in every conceivable political circumstance is beginning to look increasingly like the application of Maslow's hammer.
>
> (Banuri 1991: 25)

The preceding chapters of this book have outlined the background to, and the nature of, the neo-liberal counter-revolution in the ideology and practice of 'development' (Toye 1993). They have also described the economic, social, political and environmental impacts of SAP policies derived from them. As can be discerned from the massive number of books, articles, reports and polemics referred to in the pages of this book, SAPs and the neo-liberal ideology that informs them have been the subject of intense interest and debate for many years now. The nature of this material has been extremely diverse, relating to the different ideological standpoints of those who have considered the issues and to changes in the international context within which SAPs have been implemented and the evolving nature of the programmes themselves. This last point is important. Some of the more polemical critiques of SAPs might create the impression that the programmes have changed little over the years since their first introduction, which, as this book demonstrates, is certainly not the case. None the less, whilst it is true that SAPs have not been as monolithic as suggested by some (Herbold Green and Faber 1994), their underlying rationale remains essentially the same. This point will recur frequently in the pages which follow.

The three chapters that constitute this part of the book review a range of critiques of adjustment and the alternative proposals with which they have been associated. Various ways of organising this material could have been adopted. In the end, it was decided to separate the discussion into three broad areas.

First, chapter 7 concerns itself with what we might broadly term reformist responses to adjustment – largely relating to the 'technical' issues of national economic management and institutional capacity. In particular, attention is focused on the reactions of the IFIs to the critiques of SAPs in practice and the alternative proposals of other, more structuralist-inclined, international organisations. The central point made is that, whilst these critiques have not generally departed that far from the market-led orthodoxy, they have at least led to certain changes of emphasis and some ameliorative actions. None the less, the less hawkishly neo-liberal consensus that seems to be emerging does not, from the perspective of other more radical critics, depart significantly enough from the underlying philosophy of adjustment. Chapter 8, therefore, deals with a range of wider critiques and projected alternatives. It details proposals for radical changes in the organisation and institutional management of the international economy and the search for more profound alternatives for national economic policy-making. Chapter 9 explores the role of NGOs and social movements in the transformation of civil society in the South as an alternative to the sorts of 'social engineering' envisioned under neo-liberal restructuring.

7 Tinkering with the system
Adjusting adjustment

Ed Brown

Introduction

This chapter builds upon the major economic critiques of SAPs which have been considered in preceding chapters to consider a range of alternative models, policies and amendments which have been derived from them. As can be inferred from the chapter title, its contents are somewhat 'technical' in nature. In other words, it is essentially concerned with what we might term alternatives *within* adjustment, rather than alternatives *to* adjustment (what Felix (1998) refers to as 'Mainstream Heterodoxy' – although many of the protagonists in the debates which follow would probably object to being so categorised). Most of these critiques of SAPs start from the premise that some kind of profound adjustment to changing global economic circumstances is both necessary and desirable. Most also subscribe to, at least some, aspects of the neo-liberal critique of previous development strategies and some neo-liberal prescriptions for 'good' economic policy-making (particularly in terms of the desirability of further integration into global markets).

As such, the proposals considered here largely relate to measures that might make adjustment more successful (or perhaps less painful) and many of the issues raised have elicited some sort of positive response (however long that response might have taken) from the IFIs since the late 1980s. We begin, however, by charting attempts that have been made by individual governments to find some limited space to pursue other economic strategies or, perhaps more often, to include unorthodox elements within adjustment programmes approved by the institutions. This is followed by an overview of some of the major critiques of adjustment that have arisen from other international institutions and the alternative proposals with which they have been associated, before going into the responses of the IFIs to some of those criticisms. Finally, we consider the apparent 'post-Washington' consensus arising from the IFIs' eventual recognition of the limitations of extreme market-led approaches and their reassertion of the importance of certain types of state intervention.

Negotiating with Big Brother: conditionality or consensus?

It has already been noted that the IFIs' interpretation of the roots of the economic crisis of the late 1970s and early 1980s prioritised, almost exclusively, perceived inadequacies in domestic policy prescription and, more specifically, what was interpreted as governmental mismanagement in the Third World. Furthermore, they interpreted the ability of East Asian economies to escape the worst effects of that crisis as reflecting those governments' individual adoption of pro-trade liberalisation policies. Despite this stress on individual responsibility (and the institutions' avowal of case-by-case individual negotiation), the IFIs have attempted to ensure that governments adopt 'more appropriate' economic policies through the imposition of more or less blanket conditions on access to their financial resources. These conditions are composed of (1) 'preconditions' which must be met before any loan is agreed to and (2) 'trigger actions' which, when complied with, allow governments continuing access to the next tranche of the agreed credit (Killick *et al.* 1998: 6–7). These conditions, and the declining availability of alternative sources of finance, have meant that there has been precious little leeway for experimentation with heterodox policies at odds with the IFIs' prescriptions. Nevertheless, over the years, some attempts have been made by individual governments to follow a somewhat more independent path. Such strategies have involved 'going it alone' (attempting to stabilise and adjust their economies without recourse to the institutions' resources – hence avoiding the conditions attached) or successfully negotiating the inclusion of heterodox elements to economic policy in their dealings with the institutions.

During the early years of the debt crisis, despite the urgent need for external financial resources to confront severe economic imbalances, a small number of countries chose to 'go it alone', viewing the conditions attached to obtaining financial resources from the IFIs as bearing too high political or social costs. One example of such a strategy comes from the military regime of Mohammed Buhari in Nigeria during the mid-1980s. Buhari attempted to ease his government's foreign exchange problems, without resorting to the conditions the IMF had attached to restoring Nigerian solvency. This involved the use of countertrade via individual bargaining with a range of transnational corporations in an attempt to facilitate the exchange of Nigerian oil for commodities that, otherwise, could not have been imported. Between September 1984 and July 1985, the Buhari regime successfully negotiated deals involving the exchange of around 86 million barrels of oil with a wide range of corporations and consortia. Even after Buhari was overthrown later in 1985 and the countertrade policy had run into problems due to corruption and the fluctuation of international oil prices, the incoming government continued to pursue the policy into the following year (although this reflected a strategy of attempting to convince the population of the unavoidability of adjustment, as much as any real commitment to an alternative strategy).

Several Latin American countries (e.g. Brazil and Argentina) also attempted to stabilise their economies through unorthodox means in the mid to late 1980s. These strategies were basically an attempt to 'shock' their economies out of the inflationary spirals within which they were immersed without producing the recessive impacts associated with the orthodox stabilisation methods of the IMF. This 'shock' generally involved the freezing of prices and wages, the introduction of a new currency and the lowering of interest rates (Lehman 1993: 99). Whilst enjoying some short-term successes, in each case such strategies eventually gave way to more orthodox stabilisation methods – a fact that has either been interpreted as reflecting the inherent superiority of the neo-liberal approach (see Edwards 1995) or the lack of access to other sources of international finance that might have supported the alternative strategies. Whilst hardly outstanding success stories, these examples do at least show that some governments have attempted to pursue alternative paths in the short term, although few have been able to persist with such strategies. Generally speaking, the larger and more diversified economies have been able to get access to financial resources from a greater range of sources and, as a result, have had a little more leeway in their choice of strategy than smaller economies (Lehman 1993).

It is somewhat surprising that more attention has not been paid to the possibilities for adjusting countries to group together in order to exert more favourable terms from the IFIs. There were, in the early years of the debt crisis, some calls in Latin America (often spearheaded by Cuba's Fidel Castro) for the formation of a debtors' organisation that would threaten non-payment if there was not a more sympathetic treatment for countries suffering the ravages of economic crisis (Bello 1989: 159–160). Somewhat less radically, there were some other proposals from individual Latin American presidents to limit the level of repayment. Peruvian President Alan Garcia, for example, tried to gain wider support for his policy of setting aside a maximum of 10 per cent of export earnings to cover debt payments; whilst the Brazilian government independently suspended interest payments in 1987 (Wiarda 1987 and Bello 1989). The fact that such co-operation did not materialise, however, reflected the diverse interests of adjusting countries, the ways in which they were played off against each other and the continuing severity of financial needs. The fundamental problem, moreover, was that, given their focus upon domestic economic management rather than common external shocks, the IFIs refused to countenance negotiation on anything other than a single-country basis and severely disparaged any proposed group solutions. This was particularly ironic given the overwhelming similarity of the programmes that have been produced on this country-by-country basis.

In fact, there have been some governments that have been able to successfully secure IFI resources with considerably fewer policy conditions attached. Most successful attempts at negotiating such exceptions have, however, reflected particular strategic or temporal circumstances – particularly where the IFIs have come under pressure from some of the major Western powers to treat individual countries more leniently. Costa Rica in the mid-1980s provides one such example, where US support, given Costa Rican collaboration against its revolutionary

Nicaraguan neighbour, was used to exert more favourable conditions from the IFIs (Wiarda 1987). Similarly, Turkey's strategic importance has, on occasion, enabled it to gain access to IFI loans with much fewer pre-conditions than would have been the case for other countries (Kirkpatrick and Onis 1991); whilst Kenya certainly appears to have enjoyed a special relationship with the IMF until the late 1980s (Lehman 1993). As the years of adjustment have rolled on, however, it appears to have become more and more difficult for individual nations to resist the impulse towards neo-liberal-inspired reforms.

One further possibility in the hands of governments is that of the non-implementation of agreements once signed. Such actions (or, more correctly, inactions), some would argue, face few risks as agreements with the IFIs seem to be consistently renegotiated despite a lack of progress on individual policy requirements (Killick 1996: 223). Killick *et al.* (1998: 133) in their exhaustive study of the impacts of conditionality, for example, found that a high proportion of IMF programmes have been discontinued due to non-compliance and that World Bank SAPs have typically lasted about twice as long as originally planned due to delays in the release of tranches of funds following the non-implementation of agreed policy measures. In fact, the mainstream literature on adjustment and conditionality, even when it is relatively critical of the IFIs, depicts a bargaining 'game' where unpunished non-compliance is seen as the major explanation for the lack of demonstrable success of SAPs. Indeed, on reading some of this material one would imagine that governments are practically free to pick and choose which policy instruments they wish to implement. In their review of the South East Asian adjustment experience, for instance, Killick *et al.* (1998) suggest that when governments were 'unconvinced of the need for action, conditionality was violated, invariably with impunity' (ibid.: 136).

Logically, the conclusion reached from this type of analysis is that the apparent lack of 'punishment' for non-compliance leaves individual governments with a relatively free rein in terms of what types of economic policy they may wish to follow. This, of course, conveniently ignores the various ways in which pressure towards Western-defined adjustment is exerted upon governments – the lack of alternative sources of finance, the increasing links between IFI conditionality and access to the resources of other international and regional financial organisations (as well as bilateral aid), the need to meet pre-conditions before any adjustment process can be entered into, and so on. There is also a worrying tendency within this literature to take the underlying rationale for adjustment and the typical policy instruments required as read. Thus, the argument tends to run that since SAPs comprise a commonly accepted set of 'correct' economic policies, the reasons for the lack of successful adjustment in economic terms must be looked for in the lack of implementation of those policies. Killick *et al.* (1998: 160), for example, offer three possible explanations for the lack of economic success of adjustment. The first two, the potential role of external shocks and the undermining of the reform process through a general lack of commitment and compliance, are treated extensively; whilst the third, the possibility that SAPs themselves might be fundamentally flawed, is not even really discussed. Richard Auty (1995: 225) even goes as far as to baldly claim that 'the economic policies which can raise the per capita incomes of the developing countries to the levels of the industrial countries are now known' (for an excellent rebuttal of the case for adjustment see Fanelli *et al.* 1994).

A concern about the role of non-compliance in poor adjustment outcomes has produced a shared conviction within the mainstream literature that reform will only work effectively if individual governments have a certain level of ownership of the process – rather than being more or less forced into adjustment through IFI conditionality. Tony Killick (1996: 218–219), for example, argues that:

> The most successfully adjusting group are the East Asian 'miracle' countries but their efforts owe little or nothing to BWI (the Bretton Woods Institutions) adjustment programmes. Indeed in important and well-known ways most of them departed from BWI orthodoxies. The same local ownership appears to characterize the reform process in China and India. More tentatively, it is not clear that much of the restoration of credit-worthiness in heavily indebted Latin American countries in the early 1990s owed much to SAP conditionality. Conversely, sub-Saharan Africa was undoubtedly subjected to more conditionality per capita than any other region – and achieved the least adjustment.

This point is important in that, even when referring solely to the limited rationale of how to most successfully implement SAPs, there is a widespread (if not universal) recognition that successful adjustment must 'faithfully reflect' domestic goals and priorities. This means that, at least in principle, one would have expected there to have been some opportunities for individual governments to argue for their own nation-specific priorities within the confines of the wider approach of the institutions. In practice, however, even a World Bank study cited by Killick admits that 'local "ownership" of individual SAPs was regarded by Bank staff as "low" or "very low" in half of the programmes and "very high" in only a fifth' (Killick 1996: 219). All of this suggests that the World Bank's idea of 'ownership' of the reform process has very little real sense of local participation in it. The rhetoric, at times, paints a picture of the Bank responding to (or 'fine-tuning') locally-designed programmes, but the fundamental problem is that the Bank stubbornly refuses to believe that any reasonable consideration of the issues would produce anything but a series of proposals 90 per cent identical to their own prescriptions. Furthermore, the time taken by the Bank to produce documents and revise agreements and the limited time given to governments to respond hardly strikes of 'ownership' or 'local participation' – especially given circumstances where the Bank has sought to limit the circulation of draft documents to ten persons or less (Herbold Green and Faber 1994: 7). As Herbold Green (1998: 216) puts it:

Whether because of arrogance or naïveté, the Bank has failed to realize that 'ownership' of ideas – input from the local level, and not just the simple fine-tuning of World Bank ideas – requires consultation among a wide variety of affected actors as well as the willingness and flexibility to allow for variations on 'standard' structural adjustment.

In fact, even if the degree of local ownership of SAPs was substantially higher, it would not necessarily reflect a greater flexibility in the IFIs' dealings with borrowers but, rather, either a gradual 'softening' of the political climate within individual countries such that adjustment becomes seen as unavoidable (suggesting the need to co-operate fully with the institutions if any national priorities are to be met) or the rise to dominance of particular interest groups who see their priorities as being served by adjustment and the institutions as convenient scapegoats for the costs that will be borne by other social sectors (see Veltmeyer *et al.* 1997 for an interesting discussion of the class impacts of neo-liberal reform in Latin America).

This brings us to the glaring limitation in the mainstream treatments of the relationships between neo-liberal reform, IFI conditionality and national governments – its extremely weak treatment of the politics of the adoption of the reform agenda (returning us to issues considered previously in chapter 5). We are increasingly told that there is now a consensus that neo-liberal reforms are beneficial and necessary, reflecting an increasing commitment to market reform globally (see Edwards 1995: chapter 3, for a discussion of this consensus in Latin America). This apparent consensus would suggest an increasing local 'ownership' of the reform process, as 'realist' governments and bureaucracies are trained in 'correct' economic management – an argument which conveniently forgets the earlier assertion that the lack of positive results from adjustment was largely related to implementation slippage and a lack of commitment to the reform agenda. Analysis of internal politics within adjusting countries is routinely limited to the dismissal of opposition to the reform agenda as 'rent-seeking' behaviour from non-productive interest groups (Killick *et al.* 1998: 15), in the process effectively delegitimising the political participation of many of the poorest sectors. Meanwhile the elite sectors and bureaucrats most committed to reform are presented as somehow standing above personal interest, ignoring the substantial opportunities for personal enrichment that liberalisation (speculation) and privatisation (corruption) have afforded those able to take advantage of them (Veltmeyer *et al.* 1997: 103).

Whilst this is a simplification of a large and growing literature, it does emphasise the limitations of the mainstream analysis of the politics of adjustment and the emergence of the neo-liberal consensus. Alternative explanations for economic failure continue to be neatly sidelined. The incredible political pressure that the IFIs and Western donors are able to exert on many developing country governments is ignored and the underlying processes whereby the neo-liberal reform agenda has gradually been 'naturalised' across the globe is never addressed critically within this literature – it is simply accepted as the gradual infiltration of 'correct' economic thinking. The role of the IFIs in this process, their 'coaxing, cajoling, browbeating and threatening' of political leaders and bureaucrats into the 'ownership' of the reforms (Dasgupta 1998: 60–62) is not acknowledged.

The relationship between the 'conditioning' role of the IFIs, the embracing of market reform by particular sectors in the South and the changing nature of global economic relations warrant much greater attention than we are able to bestow upon them here. Thus, it must be stressed that, whilst we have attempted to show the inadequacies of arguments that negate the coercive elements to the role of the IFIs in the consolidation of the global hegemony of neo-liberal ideas, explanations that place sole responsibility on IFI conditionality would be equally vacuous. The shifting nature of regional and global political cultures and local class relations are also clearly important in mediating the adoption of the neo-liberal agenda. Discussions of the spread of market ideology must go beyond either the assumption that political opposition to adjustment simply reflects rent-seeking behaviour or that the spread of neo-liberalism can simply be read off from the coercive conditionality of the institutions (Veltmeyer *et al.* 1997: 34–35).

The institutional debate

Over the course of the 1980s, as the intensity and duration of the economic crisis affecting large parts of the developing world became apparent (and the social costs associated with the programmes designed to overcome the economic difficulties suspected), several other international organisations began to question the direction in which large tracts of the world were heading under the tutelage of the Bretton Woods organisations. This questioning resulted in a series of critiques of the neo-liberal agenda (often derived from structuralist or neo-structuralist academic analyses) and a range of suggested modifications in the design of SAPs or alternative policy frameworks. These interventions, given their structuralist[1] orientation, whilst accepting the overall need for adjustment suggested that policy needed to be placed within a wider and longer-term approach that was able to tackle the fundamental structural problems experienced by each region (Tarp 1993: 4). It is worth considering these critiques and the alternative policy ideas formulated from them in more detail. We will do this through a discussion of the interventions of three United Nations' organisations: UNICEF distributional critique of SAPs and the calls of the ECA and CEPAL (the UN Economic Commissions for Africa and Latin America respectively) whose calls for structural transformation and rejuvenated state intervention as an alternative to the neo-liberal adjustment policies of the IFIs.

Distributional concerns: UNICEF, SAPs and poverty

Given the record of the International Labour Organisation (ILO) in the development arena (the ILO was instrumental in bringing the basic needs agenda to the centre of the development debate in the 1970s) and its role in the protection of workers' rights (rights roundly abused under neo-liberal restructuring), it is somewhat surprising that the first important institutional critique of SAPs arose, not from that source, but rather from the United Nations' Children's Fund – UNICEF. As was explored in chapter 4, UNICEF's critique of adjustment centred upon the impact of SAPs on the conditions facing the poorest sectors in adjusting societies; issues which were originally considered purely matters of domestic policy by the IFIs and did not feature in their reports at all until 1985 (Stewart 1991). Whilst UNICEF did not blame SAPs in themselves for the existence (or even intensification) of poverty and generally recognised the case for some type of adjustment, it was argued that more could be done to protect the interests of the poor during the process (Pio 1992). This joint concern for 'promoting growth and protecting the vulnerable' led to the terminology of 'Adjustment with a Human Face' (for a wider discussion of UNICEF's role in challenging SAPs see Jolly 1990).

This position, in effect, constituted a partial restatement of the importance of the meeting of basic human needs (*à la* 1970s ILO) and the significance of 'human capital formation' to the development process (a focus that stemmed from UNICEF's central concern for children). This latter idea proposes that the provision of social services is of much more direct importance to the health of any economy than allowed for in orthodox neo-classical economics because of the impacts of poverty upon the current productive ability of large sectors of society, as well as the future potential for productive growth (Tarp 1993: 122). In effect, then, the major premise of UNICEF's proposals was that, whilst adjustment policies were necessary, they needed to be conceived as part of a longer-term development strategy that was wider than the simple pursuit of macro-economic equilibria (Tarp 1993: 126). Quite what this longer-term development strategy was and how it related to market reform was, however, left relatively unexplored – which led to some sharp criticisms of the UNICEF position.

Cameron (1992: 299), for example, argues that 'the UNICEF approach provides neither a real critique of structural adjustment as a principle nor a solution to previous problems with conceptualizing inequality and poverty'. As such, he characterises the idea of 'Adjustment with a Human Face' as basically 'palliative' in nature, in that it does not question the underlying premises of the neo-liberal approach, but merely proposes ways of assuaging its negative social impacts – suggesting that the 'Adjustment with a Human Face' focus upon poverty, whilst extremely welcome, is nonetheless insufficient and offers little beyond papering over some of the damage imposed by the programmes. He goes on to argue that 'UNICEF advocacy has been admirable for those who come within its terms of reference but offers no conceptual basis for extending the critique of structural adjustment to a wider group and deeper principles' (ibid.). This questioning of the 'targeting' principle central to UNICEF's proposals (and, incidentally, to the ways in which the World Bank has responded to the issues that they raised – see discussions below) is mirrored in discussions that have stressed the inability of such a perspective to address the root causes of poverty amongst particular social groups (Mihevc 1995: 257, for example, provides an interesting review of several critiques of the gender implications of the UNICEF position).

Finally, others have seen the whole idea of 'Adjustment with a Human Face' as fundamentally misplaced – in that, as the IFIs have picked up on the critique and taken on board its language, it has played a discursive role in tempering the possibilities for more fundamental opposition. Rist (1997: 173–174), for example, explains the idea as follows:

> here was a new way of making people believe in the harmless – even positive – character of a procedure with catastrophic effects. By a semantic trick, two opposites were joined together so that the value accorded to one was reflected upon the other, much more questionable term. A 'human face' was thus supposed to make adjustment acceptable. And with this new invention, the ideology of 'development' entered the realm of the oxymoron.

Thus, it can be argued that, whilst the UNICEF critique certainly laudably questioned the extreme economism of the initial SAP agenda and, as discussed below, produced a renewed recognition of the importance of at least addressing the issue of poverty, its acceptance of the underlying logic of adjustment has left it open to criticism in terms of its role in providing a form of social legitimacy to a basically unchanged neo-liberal rationale.

The African institutional debate: from adjustment to transformation

Following in the footsteps of the UNICEF critique of SAPs, several other international organisations have produced detailed critiques of the policies of the IFIs and a range of alternative frameworks have been proposed. In Africa, in particular, a series of critical reviews of SAPs and proposed alternatives to them have emerged from a string of meetings of African leaders and a range of publications from the international organisations working in the region. The underlying rationale of these alternative perspectives arises from the belief that:

> Africa needs fundamental change and transformation, not just adjustment. The change and transformation required are not just narrow economistic and mechanical ones. ... The need to take account of the differentiations that exist between and among societies, cultures, beliefs, attitudes and responses to external impulses as well as differences in political and economic structures has led African governments and their regional institutions to search for endogenous development paradigms that will take full cognisance of their societies, polities and economies. Whatever other criticisms one may level against the African political leadership and their regional and national institutions, like ECA and OAU, they cannot be faulted for lack of persistent effort to come out with regional strategies

and policy frameworks for meeting the challenges that confront their continent.

(Adedji 1995: 68)

In fact, comprehensive African suggestions for confronting the economic crisis date as far back as the Lagos Plan of Action of 1980. The ensuing years, characterised by a systematic lack of engagement with any of the proffered African solutions on the part of the IFIs, saw the publication of such documents as Africa's Priority Programme for Economic Recovery (1985), The UN Programme of Action for African Economic Recovery and Development (1986) and, perhaps most importantly, The African Alternative Framework to Structural Adjustment Programmes for Socio-Economic Recovery and Transformation published by the ECA (the United Nations' Economic Commission for Africa) in 1989 (Adedji 1995).

The ECA's critique of SAPs, whilst sharing UNICEF's distributional concerns and their acceptance of the need for some type of adjustment, was much wider in focus. It subjected the statistical basis of the Bank's claims for the successes of adjustment to fierce scrutiny (see Loxley and Seddon 1994). It also drew attention to the IFIs' lack of consideration of the specific circumstances facing individual nations, the worrying impacts of the adjustment regime upon national sovereignty, and, most importantly, it also cast doubt on the wisdom of several of the major SAP policy components. In relation to the latter, the ECA questioned the IFIs' assumptions about the ability of African economies to respond to price signals and suggested that, in fact, several aspects of SAPs had actively eroded the ability of African economies' productive response. The arguments underlying the advocation of privatisation and, particularly, the spurious over-generalisations about the relative levels of efficiency and corruption in the state and private sectors were also questioned, given the skewed nature of social relations in most African nations (Parfitt 1990).

The alternative policies proposed by the ECA were equally wide in nature, constituting a broad framework for reform (based around the need to overcome the structural features seen to underlie much of the economic malaise), rather than a detailed policy blueprint for successful adjustment across all nations. One of the main emphases was upon the importance of developing national capabilities and self-reliance – although this was placed within the context of adjustment to international economic realities (amidst calls for their transformation, see chapter 8 and Parfitt 1990: 137). In essence, though, the ECA's proposals constituted an attempt to combine the IFIs' concern for short-term adjustment to current economic circumstances with a renewed focus upon much more profound (and less recessive) transformations in the region's economy, social structure and political institutions that might overcome the structural problems identified as lying behind the economic crisis (Teriba 1996: 15). This position is encapsulated in the ECA's language of 'transformation' rather than 'adjustment'.

As such, the types of measures highlighted within the framework included the following possibilities (these are termed possibilities here to reflect the much less prescriptive stance of the ECA – in comparison to the IFIs – in terms of what policies might be applicable to the circumstances of individual countries):

- reduction of state spending on the military and other non-productive activities and its redirection towards social provision and targeted sectors of the economy;
- maintenance of fiscal solvency through reduced and more strategically directed use of deficit financing, the elimination of blanket subsidisation and the expansion of the tax base (as well as increased taxation of luxuries);
- promotion and selective financing of the agricultural sector (through guaranteed minimum prices, land redistribution and increasing the foreign exchange allocation for agricultural inputs);
- credit arrangements, codes of investment and interest-rate policy designed to encourage productive activities rather than speculation and capital flight;
- institutional measures to support the longer-term prospects for the strategy such as the promotion of agricultural extension services, the strengthening and reorganisation of credit provision, legislation to clarify social rights, ownership and responsibilities and greater attempts to enhance public participation in policy formation and the political system more generally.

(Parfitt 1990: 137–138; Tarp 1993: 144–145)

The ECA's proposals certainly departed from neo-liberal orthodoxy in various ways, particularly in terms of the longer time-scales involved, the wider range of issues tackled, the direct concern for social provision and the promotion of (albeit reconceptualised) state intervention. At the time, there was a vitriolic series of exchanges between the World Bank and the ECA over the interpretation of the events of the previous decade and appropriate strategies for the future. The IFIs' initial response to the ECA's proposals was to condemn them for, in reality, representing little more than a renewed call for the heavy state intervention of the past. Relatively rapidly, however, a more conciliatory approach was adopted which stressed the factors held in common and, gradually, over the intervening years a range of the issues raised by the ECA have been assimilated into the IFIs' position on the longer-term transformation of the region – although the policy frameworks suggested for achieving those goals have changed little. As Mihevc (1995: 117–118) has noted, this could be interpreted as an attempt to co-opt the ECA's critique by suggesting that there was a general agreement over the long-term goals for the region and many of the specific critiques raised by the ECA but that these could only be adequately addressed by the continued, if deepened, application of SAPs.

Elsewhere, the ECA's proposals were generally well received, although there were criticisms of the lack of political analysis reflecting on how the desired transformations were likely to be achieved. Thus, Teriba (1996: 19) suggests that the ECA's strategy, whilst mindful of the need for political realism (and particularly the need for the expansion of meaningful participation in the political system), remained somewhat all encompassing and it was never clear quite how the proposed transformations (i.e. the more radical longer-term intentions) were to be achieved. Furthermore, leaving aside the question of the political will of African leaders to implement some of the more far-reaching proposals for redistribution and enhanced participation (let alone the autonomous role assumed of the state in guiding those transformations), the ECA's position also failed to sufficiently question the capacity of the state to fulfil the central, albeit reconceptualised, role it was to play within the economy (Carmody 1998: 33 and Tarp 1993: 146).

Neo-structuralism in Latin America: CEPAL

CEPAL (also referred to as ECLA – its English acronym), the Latin American equivalent to the ECA, has also played a significant role in exploring alternative policy frameworks to the neo-liberal focus of the IFIs. Lampooned and side-lined during the 1980s as the organisation responsible for the import-substitution (ISI) framework so despised by neo-liberals, recent years have seen CEPAL articulate a critique of neo-liberal adjustment which bears some resemblance to the ECA's ideas. CEPAL's contribution can be encapsulated under the rather unwieldy term of neo-structuralism – reflecting CEPAL's historical association with the structuralist pole of the long-running confrontation between neo-classical and structuralist approaches towards the Latin American economy. The discussion which follows places CEPAL's position within the context of the gradual re-emergence of structuralist thought from its 'wilderness' years.

Neo-structuralist reactions to neo-liberal adjustment have generally involved combining a recognition of the need for adjustment (and the validity of many neo-liberal prescriptions for 'sound' economic management) with calls for a renewed form of state intervention in order to tackle the specific structural problems seen as underlying the region's economic problems (problems that SAPs had exacerbated, or, at best, ignored). CEPAL's particular contribution has been to focus upon inequality as the most important and intractable structural problem afflicting the region. Their analysis has been based around a twin recognition of (1) the ways in which the state-led 'populist' strategies that dominated Latin American economies in the post-war years paid insufficient attention to engendering sustainable economic growth (and also led to their creation of unwieldy inefficient state sectors) and (2) the lack of attention that is now being paid, despite the rhetoric of democracy and opportunity, to issues of equity and equality of opportunity within the dominant neo-liberal model of the contemporary Latin American scene. To bridge the gap between the inefficiencies of the past and the inequality of the present, CEPAL argues, will involve combining a concern for equity (which is seen as necessarily including land reform and redistribution of wealth, as well as the targeting of deprived groups) with an appreciation of some of the lessons concerning economic efficiency derived from the neo-liberal critique of the more state-centred and inward-looking trajectories of the past (see Ramirez 1993 for a wider discussion of neo-structuralist alternatives to adjustment). Thus, for example, the traditional structuralist belief in the centrality of the industrial sector and the need for the state to actively promote industrialisation (as articulated in the ISI strategy) has not been abandoned but, instead, reinterpreted through a recognition of the importance of the export sector to that process (Martinussen 1997: 78).

Similarly to the ECA, the Latin American neo-structuralist proposals are based around a reoriented state intervention as the key to a more sustainable economic model. This does not, of course, mean a return to the 'statism' of the past (with its productive role and top-heavy bureaucracy) but, rather, a redefined slimline state, one that intervenes more actively than has generally been allowed for within neo-liberal thinking. As Duncan Green argues in relation to CEPAL, this

> does not mean a return to the octopus state of the 1960s. CEPAL accepts that production should wherever possible be left in the hands of the private sector. Instead it argues for a managerial state, in alliance with and regulating the private sector, intervening in the economy to move it towards higher levels of technology and industrialization, training and caring for its population.
>
> (Green 1995: 189)

As such, neo-structuralist analyses have called for a more sequential and gradual approach to adjustment (thereby avoiding its worst recessive impacts), a shift (in the context of continued efficiency drives) in government spending towards infrastructure, health and education, the active promotion of more labour-intensive sources of employment in agriculture and industry (through the funding of agricultural programmes, export diversification, etc.) and also a fundamental reappraisal of the region's recessive taxation system (Ramirez 1993: 1,025–1,033).

It is unclear, however, quite how far the neo-structuralist school represents an alternative to the neo-liberal model. Whilst there is a renewed emphasis on the role of the state in development and a concern for equity and participation, some have suggested that it merely represents a slightly more widely-defined version of 'Adjustment with a Human Face' and appears to offer little in the way of more radical departures from the underlying market-led model. Indeed, many commentators, sympathetic to the neo-liberal agenda, have interpreted recent neo-structuralist writings as, more often than not, broadly supportive of the direction of economic policy in the region under the neo-liberal reforms of the past fifteen years or so, rather than a challenge to them. Zuvekas (1997:157), for example, approvingly suggests that 'structuralism, along with its focus on inward-oriented development, has been transformed into a neo-structuralism that accepts much of the neo-liberal agenda (including macro-economic balances, price liberalization and external competitiveness)'.

This probably underestimates the differences between the neo-structuralist and neo-liberal approaches (although, as we shall see below, they have certainly been moving closer together during the latter part of the 1990s) and, in particular, fails to recognise CEPAL's emphases upon broader issues such as redistribution, democratisation and equity. Nevertheless, many questions remain regarding CEPAL's ability to adequately address those issues. These revolve around similar issues to those discussed regarding the ECA – how, for example, will the inefficient, authoritarian and often corrupt state sector in Latin America be transformed into the efficient and trustworthy regulator of the economy envisaged in neo-structuralist analyses? Furthermore, in relation to the key issues of equality and redistribution, how will the powerful beneficiaries of the inegalitarian strategies of the past (and, perhaps even more importantly, the present) be persuaded to peacefully hand over power and advantages without simply transferring their financial resources out of the country? (Green 1995: 190). CEPAL's vision, much like the ECA's, is one of an Asian-style technocratic state that is able to stand above political interests, but such a strategy rests uneasily with CEPAL's concern for increased political participation and decentralisation. In some senses, this contradiction recreates the problems with earlier structuralist approaches to development within which the state was unproblematically heralded as the answer to market failure without addressing 'both the institutional and the political requirements of effective state intervention [and] the possibility that in the absence of the required institutional and political bases, which provide a high degree of organisational strength as well as autonomy from sectional interest groups for the state, state intervention would prove to be counterproductive' (Onis 1995: 99).

Whilst there are obvious differences in the critiques and proposals of the various institutions that have been considered here, some common elements emerge within structuralist economic analysis. First, most accept substantial parts of the neo-liberal message, particularly through recognition of the profound changes that have occurred within the global economy. Second, much structuralist analysis revolves around the need for political reforms that will strengthen the state, whilst also making it more accountable, and, therefore, more effective in its developmental role; this tendency can be sharply contrasted with the overall neo-liberal antagonism towards the state. Third, one further element of rejuvenated state involvement, not directly considered above, involves the widespread support for measures designed to tackle the social consequences of adjustment. Most proposals have, therefore, called for policies directly designed to tackle poverty through the selective use of subsidies. Finally, much is made of the need for greater sensitivity to individual needs and regional diversity.

World Bank reactions to the structuralist critique

> The Bank's research department in the early 1980s responded to unorthodox thought defensively. To parody only slightly: 'We cannot be sure whether unorthodox thought A is true or not. If it is false it is dangerous, but if it is true, it is even more dangerous'.
>
> (Mosley *et al.* 1991: xxiv)

Structural adjustment in 1994 is by no means what it was in 1981 either in terms of conceptualization and dialogue or of operational guidelines and praxis. As neither states nor – especially – the World Bank like to admit own error openly (pointing to other people's is something else), but states and – again especially – the World Bank do learn from and respond to experience, the shifts are hard to follow for one fully involved in the process and near impossible to grasp in detail – or overall scope – for an outsider.

(Herbold Green and Faber 1994: 5)

Over recent years the World Bank (and, to a lesser degree, the IMF – although the Asian crisis appears to have prompted some more profound internal dissent even here) has taken on board some of the structuralist critique considered above and the nature of adjustment programmes has certainly been transformed substantially. At the most minor level, the Bank has responded to specific technical critiques to do with the time-scales involved, the sequencing of reforms and the applicability of specific policy measures (see Greenaway and Morrissey 1993; Killick and Stevens 1991; Kirkpatrick and Weiss 1995; Mosley *et al.* 1995; and Herbold Green 1998). More generally, since the late 1980s the World Bank has at least paid lip service to the structural concerns of its critics. Relatively rapidly, it was realised that SAPs were not going to have the desired economic impacts as quickly as had been hoped. As such, the IFIs gradually began to come round to the view that there were some important structural features at play; although this seems to have reflected a search for structural causes and unforeseen processes through which the non-achievement of adjustment goals might be explained, rather than any doubts about the overall applicability of neo-liberal economic doctrine (Gibbon 1996: 758).

A new focus on poverty?

Towards the end of the 1980s, in response to the mounting criticism of the social costs of adjustment and political unrest from those adversely affected, the IFIs finally began to address the issue of poverty much more directly, as it became clear that adjustment was going to be a much longer process than had originally been envisaged and one which would require some special measures to assist the poor (Mihevc 1995: 125 and Cameron 1992: 294). By 1989, even the IMF had stated that 'it would be morally, politically and economically unacceptable to wait for resumed growth alone to reduce poverty' (IMF Survey 3 April 1989), whilst the following year the World Bank were to devote their annual World Development Report specifically to tackling poverty.

The anti-poverty strategy outlined by the Bank in the 1990 report consisted of (1) the advocacy of broadly based 'labour-intensive growth based on appropriate market incentives, physical infrastructure, institutions and technological innovation', designed to generate jobs and income for the poor, and (2) social investment to improve poor people's access to, and the 'adequate provision of, social services, including primary education, basic health care, and family planning services'. In addition, the report argued for transfer payments to 'those who would not otherwise benefit' (from the two-fold strategy just outlined) and 'safety nets' to protect those most vulnerable to 'income-reducing shocks' (World Bank 1990: 138). This new anti-poverty rhetoric led, over the next few years, to the development of Poverty Assessments (PAs) which were to feed into the existing mechanisms employed by the Bank in their dealings with individual adjusting countries (see World Bank 1992a) and the creation, in 1993, of a vice-presidency with specific responsibilities for human development and poverty reduction (Killick 1995: 321).

Some have interpreted these new developments as a relatively profound transformation in the adjustment process. Toye and Jackson (1996), for example, whilst critical of the rather limited conceptual basis of the Bank's understanding of poverty, are heartened by the transformation in the Bank's approach, suggesting that:

> During the presidency of Lewis Preston, the Bank has transformed itself from a tardy follower (or sometimes outright critic) of the poverty agenda into a clear leader of important initiatives being taken world-wide to combat both long-term structural poverty and the conjunctural poverty which arises from public sector restructuring and other adverse effects of structural adjustment policies.
>
> (Toye and Jackson 1996: 66)

Others have, however, been far more critical of how this anti-poverty strategy has been implemented in practice. Toye and Jackson's delineation of 'structural' and 'conjunctural' poverty is a good place to start when considering these criticisms. Much has been made of the Bank's failure to integrate a concern for both types of poverty into the formulation of SAPs. The Bank claim to address both: structural poverty through the overall focus on labour-intensive economic growth and the removal of the impediments to economic activity for all sectors and specifically SAP-induced poverty through targeted programmes. Specific critiques frequently relate to studies of the latter – the Bank's targeted poverty alleviation programmes, job creation schemes, community improvements, and so on. These are specific projects designed to meet particular needs amongst those most severely negatively affected by adjustment. As such, they are designed to deal with what are perceived as the short-term effects of adjustment upon specific groups. Such schemes have included providing training, credit, agricultural land or severance payments to those made redundant (Stewart 1991: 1,858) and, by and large, have been funded through the redistribution of resources within the social sector's budget rather than any overall increase in social spending (Mihevc 1995: 124).

There have also been a few attempts at more integrated projects. Perhaps the most influential scheme was Bolivia's Emergency Social Fund, first implemented in late 1985, which then gained World Bank support in 1987. It was a four-year programme designed to provide short-term employment and basic social services during a period of severe adjustment; although it was particularly innovative in that it was demand-driven (Graham 1992: 1,234). Whilst this scheme (and others like it) was more comprehensive in focus, it was still basically an 'add on' to the design of the adjustment process, rather than an integral part of overall strategy. This is, then, the major critique of such initiatives, that a concern for poverty and inequality has not been integrated into the overall rationale of adjustment, being instead relegated to the development of these add-on programmes (Tarp 1993: 146). The response of the Bank to such criticism would be that it is unfair to criticise the programmes on such a basis since they have been specifically designed to deal with the conjunctural poverty produced through the adjustment process itself (an innovation called for by many of its critics), whilst it is the overall focus on labour-intensive economic growth and the wider changes beneficial to the poor that are part and parcel of the adjustment process that will gradually tackle the structural causes of poverty.

Needless to say, such assumptions have also themselves been challenged. The advocation of, admittedly now labour-intensive, economic growth as a strategy for dealing with structural poverty deviates little from the Bank's original line that economic growth and the removal of impediments to individual economic freedom imposed by the state would work to the benefit of the poor. Whilst the new labour-intensive slant to the focus on economic growth has the laudable goal of decreasing unemployment, it has, none the less, been accompanied by the broader neo-liberal focus upon driving wages down (let alone the lack of attention to the protection of basic workers' rights), thereby exacerbating economic insecurity and inequality (Oxfam 1994 – cited in Green 1995: 55).

Overall, whilst there has been a shift in the Bank's rhetoric and some recognition of the problematic social impacts of SAPs, poverty (and, indeed, inequality) still does not figure in the overall rationale for adjustment within the Bank's approach (Toye and Jackson 1996: 58–59). Furthermore, with the partial exception of basic services and emergency food security, the lack of articulation between the new poverty focus and the neo-liberal core of adjustment has meant that the former has remained underfunded and undervalued (Herbold Green 1998: 221). Thus, it has 'only tangentially and indirectly permitted inequality and poverty to creep back onto the international conceptual and policy agendas' (Cameron 1992: 296).

Some are even more suspicious, seeing the more effective targeting of the poorest sectors inherent in the new approach as shorthand for reducing overall social spending budgets and politically 'managing poverty' at low cost. To Chossudovsky (1997: 66–67) 'the SEF (Social Emergency Fund) officially sanctions the withdrawal of the state from the social sectors and "the management of poverty" (at the micro-social level) by separate and parallel organisational structures. ... A meagre survival to local-level communities is ensured whilst at the same time the risk of social upheaval is contained.' In essence, whilst the schemes are beneficial to the few that gain from them, they can often act as a smokescreen for the more decisive tendency – the continued decline in social spending and the retreat of the state (Green 1995: 103).

Despite the limitations in the IFIs' shifting approaches towards poverty alleviation, they are certainly an advance on the narrow economism of the earliest SAPs. The attention that has been devoted to poverty reduction has raised some important issues and starkly revealed the limitations of the earlier excessive focus on market liberalisation and allocative efficiency – with most commentators now accepting the importance of tackling structural poverty. Given this, much recent literature has revolved around the methods that might be employed to ensure widespread support for anti-poverty programmes from national governments, local elites, etc., as well as the international institutions. The question remains, however, over who has the ultimate responsibility for tackling poverty. Does the Bank's new focus on poverty reduction lead to further impositions on national sovereignty, or is it an opportunity for the international community to pressurise governments into meeting the needs of their poorest citizens, given the dismal record of many governments in that regard? There is no doubt that this is a tricky issue. One is torn between a recognition of the self-important, Western-centric and often arrogant way in which the international institutions depict the governments that they work with (and the serious lack of alternatives that many national policy-makers have at their disposal) and yet, at the same time, it is possible to sympathise with a Bank position that suggests that:

> many governments are reluctant to define a poverty agenda ... in order to avoid sacrificing political patronage. Instead they use rhetoric such as 'everyone is poor' to avoid making a political commitment to poverty reduction that would be unpopular with those who keep them in power.
>
> (Addison 1993: 2 cited in Killick 1995: 318)

Nevertheless, one must remain critically aware that such interpretations have been used to justify the institutions' previous lack of attention to poverty as a, so-called, politically sensitive issue – reflected in the old IMF position that it was 'only governments' that should concern themselves with the distributional impact of economic reform and not the institutions.

Reconsidering the role of the state: good governance and market-friendly intervention

As we saw in chapter 5, the World Bank has also gradually been responding to criticisms of its over-dependence upon the market and its attack on the state, through the emergence of new positions that appear to represent a movement away from neo-liberal state minimalism. This renewed interest in the state can be encompassed under the Bank's twin concerns for what it now terms 'good governance' and 'market-friendly' state intervention (Kiely 1998). The latter represents a recognition of the need for states to have a more direct role in economic development and, perhaps, more than anything, reflects how the

Bank has grudgingly come to recognise the validity of alternative explanations of East Asian economic success. The Bank suggests, however, that, whilst the state has clearly played a wider role within East Asian economies than it had previously argued, that intervention was 'market friendly' rather than 'market distorting' (World Bank 1993a). In essence, what this means is that, in the Bank's view, where the state intervened in East Asia it did so in a very limited way or operated policies that counter-balanced each other.

On the basis of this analysis, the Bank argues that interventions should take the form of checks and balances, rather than wider measures that might distort market signals (Kiely 1998: 67–69). Ray Kiely provides a robust rebuttal of these claims (stressing a much wider role for intervention within Asian development strategies than allowed for by the Bank) and emphasises how, on closer inspection, the Bank's rediscovery of the state does not depart in any significant way from the premises of its earlier pronouncements. Kwon (1994: 635) interprets this as reflecting 'almost a textbook example of neo-classicists visibly confused but too proud to admit their failure – having been so quick to blame government for economic failures in the past, they are now reluctant to admit a positive role for government in a successful economy' (ibid.: 635). Nonetheless, such an interpretation misses the consistencies within the Bank's position in that, whilst there is now a recognition of the important role that the state can play within the economy, that role is exclusively viewed in terms of its ability to create the appropriate conditions for accelerating and deepening a continued process of market liberalisation. As such, the rejection of a minimalist state does not necessarily represent a reversal in World Bank policy but perhaps astute political manoeuvring (Hildyard and Wilks 1998: 50). As Ben Fine argues:

> given the previous stance in favour of state minimalism, even if serving as a veil for considerable discretionary intervention in practice, there has been a problem in addressing what role the state should play given its continuing importance. One cannot argue that the state should do nothing but also debate what the state should do. The World Bank has been disarmed by its own ideology. Now, rather than becoming state-friendly, a more appropriate interpretation may be that it is seeking to be more influential than before over what the state does – both in depth and scope. ... In short ... [this] allows the World Bank to broaden its agenda, whilst retaining continuity with most of its practices and prejudices.
>
> (Fine 1999: 11)

Whilst the World Bank may not have shifted very far in relation to its acceptance of state intervention within the economy, recent years have seen considerable attention paid to the institutional environment and issues of governance (Gibbon 1996: 764). By 1993, the World Bank had moved beyond a simple 'hatchet-job' on the role of the state in development failures to propose a new vision of what a 'good, efficient' state should do in order to better implement adjustment – encapsulated under the idea of 'good governance'. The World

Bank (1993b), for example, contains a list of elements of 'poor governance' said to underlie many examples of failed adjustment programmes (factors mentioned include the unaccountable nature of bureaucracies, the arbitrary nature of decision-making, unjust legal systems, the exclusion of civil society from the political process and corruption and the abuse of power), as well as an outline of what 'good governance' might (and should) involve – fundamentally a strong, and yet sharply delimited, state (Baylies 1995: 326).

To a certain degree, this new focus upon 'good governance' represents a recognition of the importance of political context to economic policy and a partial acceptance of some of the arguments concerning how SAPs may in themselves have actually sometimes hindered the ability of the state to administer any sort of economic reform. Nevertheless, as Kiely (1998: 68) argues, the Bank's embracing of 'good governance' is, in essence, 'a means to an end'. By this, he means that the Bank's concern for the institutional structures of government and its rediscovery of democracy have little to do with any reflection on the nature of the state and its biases or a desire to increase participation and consolidate democratic political frameworks for their own sake, but rather are limited to the search for 'efficiency' and 'sound development management'. As he goes on to explain, in contrast to earlier Bank positions, such management is now seen as requiring, not just less government, but better government – government that is devoted, not to direct intervention, but rather to creating the appropriate conditions for successful economic transformation within the private sector (Kiely 1998: 68). It illustrates the Bank's continued antagonism towards the state, which is still seen as an evil, albeit increasingly recognised as, in some form, a necessary one. Thus, in the final analysis, the constraints imposed upon successful economic reform are still seen to lie in the inefficiencies of the state, whilst private sector activity is still seen as unproblematically beneficial. There are precious few examples where the Bank has conceded that private sector characteristics (monopolistic, corrupt, inefficient or speculative) can act against the successful outcome of adjustment (Gibbon 1996: 776). Such issues are excellently summarised by Lance Taylor:

> In countries undergoing adjustment, instead of disappearing under market-friendly policies, corruption has surged over recent years, spawned by export incentives, speculative urban finance, privatization and stock exchange operations and fiscal incentives. ... Such social developments are beyond the ken of the Washington model, which cannot absorb the fact that rents and corruption often rise instead of declining when old forms of market regulation are suppressed.
>
> (Taylor 1997: 149)

This discussion brings us into the thorny question of the relationship between governance (as conceived by the Bank), democratisation and economic development. Despite the one-to-one relationships now being postulated by the Bank between institutional efficiency, political liberalism and effective economic outcomes, such linkages have not long been established in the literature and are by no means universally accepted. In the 1970s and 1980s, neo-liberal economic reformers often secretly (and sometimes openly) saw authoritarian governments as being more able to deliver policy reform than democratic ones (Nelson 1992: 310). The continued failures of adjustment led to the new focus upon institutional efficiency and the rather uneasy advocacy of democratic reforms as being necessary for economic reform (although the argument is always couched in these terms rather than in terms of the desirability of democratic reforms for their own sake). How might we interpret these changes? To some critics the new focus on governability is to be lauded as a final recognition by the IFIs of the importance of the state to the successful implementation of economic reforms. To some neo-liberals it is problematic in that, despite the recent turn arounds in Bank policy, they still believe that successful economic adjustment requires strong authoritarian control which is something that meaningful democratic participation cannot provide. To others, it represents a new tranche of policy conditionality and a further attack on the sovereignty of individual nation-states; whilst still others see it as embracing a very limited idea of democracy, in that, despite the rhetoric about building up the organs of civil society, good governance focuses exclusively upon the 'effectiveness' of the state (Baylies 1995) – although this interpretation is at odds with the Bank's evolving relationship with the NGO sector – an issue returned to in chapter 9.

Conclusions: the post-Washington consensus?

The institutions have referred to the changing emphasis and scope of adjustment programmes as a 'learning process', whereby the experiences of successful and unsuccessful reforms have refined their understanding of the adjustment process and reinforced the importance of a wider range of elements in its successful completion. As Engberg-Pedersen *et al.* (1996: 17–18) suggest, however:

> an alternative way of reading the same process is that many IFI assumptions about economic and social actors in recipient countries were naive and ideologically based. Subsequently there has been less of a learning process than a combination of ad-hoc rectifications and a series of uncoordinated and ill-thought-out plunges into new areas which remain poorly understood.

Whilst this cynical view of the changing rhetoric and concerns of the Bank can certainly be applied to the 'rediscovery' of poverty in the early 1990s, it does not quite capture the shifts in attitude towards the state that have occurred towards the end of the decade. The publication of the 1997 World Development Report, in particular, seems to have marked something of a culmination in the gradual 'meeting of minds' within mainstream thinking on adjustment since the first widening of the policy debate in the early 1990s

(Herbold Green 1998; Fine 1999; Pugh 1997; Cornia 1998; Moore 1998; Hildyard and Wilks 1998 and Martinussen 1998). That report's focus upon the need for 'effective' state intervention and its conciliatory tone towards alternative positions, together with the questioning of the previous neo-liberal consensus emanating from Joe Stiglitz (the new senior Vice-President and Head of Economic Research of the Bank) through his insistence on the prevalence of market imperfections in Third World settings, have prompted some observers (including Stiglitz himself) to talk of a new 'post-Washington' consensus (Fine 1999) – a consensus based on an apparent rapprochement between a neo-liberalism now cognisant of the importance of institutional context and a critical structuralist or institutionalist economics that has accepted the case for many aspects of neo-liberal reform.

Some critics have taken great heart from these changes and have stressed that, whilst there may not be a consensus on the types of policies and reforms necessary for development, the Bank's new openness does suggest at least a more healthy climate for meaningful dialogue amongst policy-makers and mainstream economists (and even many NGOs). Mick Moore suggests that the 1997 World Development Report, for instance, provides for the possibility of a reasonable working consensus,

> not so much by providing specific policy conclusions on which we can all agree ... but rather by providing a language, a perspective, and a set of concepts and concerns which will make it easier for people with divergent views actually to debate and discuss with one another in a productive way.
>
> (Moore 1998: 39–40)

Whilst such views of the changing position of the Bank are somewhat idealistic (especially given the Bank's record on dealing with dissent and the many consistencies that remain within their policy frameworks), they do raise some important issues for the Bank's critics. First, at least in terms of its general policy statements, the Bank has shown some responsiveness to criticism (as we have repeatedly tried to stress within this book, SAPs in the late 1990s are very different to those of the mid-1980s). Second, whilst there may be disagreement on the motivations underlying those shifts in policy, and how far they represent a departure from the underlying neo-liberal philosophy of adjustment, any recognition of the importance of tackling poverty, the centrality of political participation and institutional accountability, whatever their limitations, is surely to be welcomed. Third, even if the move towards the positions of its critics is more apparent than real, any advance in the openness of the Bank to criticism and the possibility for more meaningful dialogue is also an important step, even if it does not necessarily represent the new consensus postulated.

Nonetheless, whilst there is little doubt that the neo-liberal triumphalism of the 1980s is now long passed and that the pronouncements of the Bank now reflect a much more diverse set of influences, the questions remain as to how much has really changed and with what likely impacts.

First, given past experience, it is worryingly premature to assume a one-to-one relationship between policy discussions and the day-to-day operational practice of the World Bank (Cornia 1998: 32). To take a more historical example, despite the repeated commitments to making poverty reduction an integral part of adjustment and the way in which the World Bank has repeatedly stressed its role in such activities as part of its *raison d'être*, such concerns have still tended to be marginalised within the operational and policy framework of the organisation. Herbold Green and Faber (1994: 7–8) illustrate effectively the limitations of changes in Bank philosophy during the earlier part of the decade:

> Conceptually the Bank has ... resurrected Robert McNamara's war on absolute poverty. ... Operationally, however, it still marginalizes it into parallel projects even if a state wishes to integrate it into its macro strategic policy and resource allocation core. ... Gender and ecology are even more tack-ons – even if the Bank has a serious unit in the latter field. In principle the agenda is more or less complete – distribution, poverty, gender, ecology, accountability, participation, a pro-active (not a passive) state enabling role are all on it. But all are in practice marginalised or at least ill-integrated with main sectoral, let alone macro, strategies. Indeed the dreaded Public Expenditure Programme missions sometimes appear to ride roughshod over even major sectoral issues and proposals of the Bank's own sectoral missions.

Whilst the internal shifts of the Bank have probably deepened since that opinion was expressed in the mid-1990s, it does suggest the importance of continuing internal struggles within the Bank over how far its orthodox neo-liberal focus has been, or needs to be, superseded. The possibility remains that the attitudes expressed within the 1997 World Development Report (and by the Vice-President) do not hold sway across the Bank and may not, therefore, have a significant impact upon the evolution of Bank activities. Furthermore, there is also the possibility that the apparent openness of the Bank to change is a rhetorical device designed to offer the cosmetic semblance of change, whilst the (market-reforming) business continues as normal.

Second, even if the new attitudes towards the state were to take root strongly within the operational practice of the Bank, they may not represent a substantial change in the underlying rationale of adjustment, but rather its redimensioning (Felix 1998: 212). Support for such a view comes from the observation that there are far more continuities with previously articulated Bank positions than might be suggested from its recent pronouncements. The changes in emphasis have not challenged the over-riding economism of the Bank, nor its continued assumption that the correct economic and social policies are known and people must be persuaded to follow them (once again a 'correct' form of state intervention is being proposed, rather than a wider discussion of alternative forms of state action and organisation – Martinussen 1998: 70). As such, the attention being paid to issues of governance and intervention by the state to correct market distortions represents, not a substantial redefinition of adjustment, but rather a further re-evaluation of the time-frames involved (Herbold Green 1998: 218).

Seen in such a way, the whole reform process of recent years can be seen as the gradual encroachment of the IFIs into more and more areas of national life. The response of the IFIs to each wave of criticism has generally been that of expansion – through the creation of new units (on the environment, on poverty, on women, etc.) that have often employed some of the most vehement of their critics – in the process turning their energies towards internal reformist battles within the institutions (Green 1995: 53). In effect, the gradual tacking on of more and more areas of national life into the hitlist of factors to be adjusted has simply resulted in ever more aspects of political life within adjusting countries being taken out of the hands of national governments (Engberg-Pedersen *et al.* 1996: 17 and Green 1995: 54). In this light, the new consensus being pursued by the Bank would appear to encompass (1) widespread agreement on the types of economic policies that will lead to economic growth – broadly in accordance with the SAP model, (2) the recognition that there are a range of more structural features that have tended to work against the successful adoption of such policies and that, therefore, the process of adjustment will be a long and drawn-out process (during which the most vulnerable sectors will need to be protected) and (3) the need for a range of interventions that will, over time, create the appropriate conditions for successful economic adjustment – through the removal of the structural and institutional impediments to the spread of market reform and the wholesale transformation of social and cultural norms.

As such, as Hildyard and Wilks (1998: 51) argue, 'the Bank's sole benchmark for assessing the "effectiveness" of political processes, procedures and institutions is whether or not they act as lubricants or potential barriers to deregulation and fiscal discipline'. Thus, the new focus on a much wider set of social, political and cultural transformations, as part of a continued process of neo-liberal reform, could be seen as moving us into the realms of social engineering – through the long-term moulding of the institutional structure of whole societies to the formal (and limited) goals of good governance and administerial competence whilst, at the same time, utilising the transformation of civil society (and in particular the international NGO sector) as protagonists of a new market-oriented (or perhaps market-friendly) society. This interpretation of the shifting attitudes of the World Bank raises some important questions which suggest that the new, wider, rationale for adjustment may turn out to be more misguided and impositional than the econocentric model of the 1980s. As Ben Fine (1999: 13) puts it:

> (the new consensus) opens up an agenda for those who opposed the old consensus, but there is an admission price in terms of accepting the social as based on microfoundations and capital as based on market or non-market imperfections. Notably absent will be a political economy based on class and power and capital interpreted as a social relation rather than as a

non-physical atomized resource. In short, where the developmental state literature previously stood as a critique of the old consensus, it can now either be overlooked or be repackaged as new in terms of a much less radical content attached to market imperfections and social capital.

In effect, what this suggests is that the terms of any emerging consensus, on the lines expressed by the Bank, run the risk of depoliticising the state, of making the whole question of the choice of economic strategy and forms of political participation a matter for appropriately trained experts. As Kiely (1998: 74) suggests, by 'focusing on governance, the Bank has reduced the politics of development to a purely technocratic issue' (ibid.: 74). What this means is that, through its interpretation of political and cultural context in terms of market imperfections and the construction of social capital, the World Bank has reduced political analysis to issues of institutional 'efficiency'. There is no discussion of the underlying factors that shape the different ways in which states act or, indeed, of the non-economic roles of the state (Martinussen 1998: 69). Neither is there any recognition of the 'extent to which the causes and remedies of poor governance in developing and transitional countries lie outside their own frontiers' (Moore 1998: 40) nor any discussion of 'to what ends and in whose interests the state operates' (Hildyard and Wilks 1998: 50).

This final point is important since it cuts right to the centre of the mainstream debate on development which looks set to be dominated by the new consensus over the coming years. Whilst the preceding discussion has pointed out the limitations of the World Bank's shifts in perspective over the course of the past decade, many of the objections raised concerning the limited attention paid to (or the weak conceptual framework for understanding) the politics of state intervention and democratisation, could equally have been applied to the structuralist alternatives considered above. The apparent shifts in Bank policy and the new rhetoric about institutional efficiency, state intervention and poverty alleviation have actually embraced many (although certainly not all) of the issues raised in structuralist writings and make the agenda of the new consensus attractive to those who share the Bank's formalistic vision of politics, even if they continue to question the economic underpinning. Neo-liberal perspectives on economic policy and society are increasingly taking the form of unstated assumptions that continue to dictate the terms of the international debate on development. As such, the vast majority of ideas considered within this chapter have tended to be limited by those assumptions. The following chapters take the debate a stage further through an examination of some more radical departures from the neo-liberal model.

Note

1   We have termed these critiques structuralist, reflecting the particular economic orientation which underlies much of the material. Not all of the sources reviewed here quite fit into that pigeon hole, but there are enough common factors for us to talk of a loose structuralist approach to adjustment. Tarp (1993: 127) identifies three

strands to structuralist critiques of SAPs. First, what he terms neo-classical structuralism. Here, most of the neo-liberal emphasis on market efficiency is conceded but it is argued that, in reality, markets continue to be partial and distorted in many 'developing' country settings. As such, markets may not respond to price changes in the ways predicted by neo-classical economists due to supply-side constraints. This, they argue, provides a justification for 'non-price rationing' and other state interventions. A second position distances itself further from the neo-liberal position, through more fundamental concerns about the distributional effects of adjustment, the 'flexibility' of developing economies in responding to price signals and the need for attention to be directed more specifically towards productive transformation. Finally, the third position sees macro-economic disequilibria as having their roots in the underlying structures of society rather than purely economic phenomena (ibid.: 127–130).

8 The search for more radical alternatives

Ed Brown

Introduction

This chapter seeks to manoeuvre the terms of the debate beyond the search for modifications within the ascendant neo-liberal policy agenda (or for 'spaces' within the international economic system as presently constituted) to consider:

1. perspectives that view fundamental transformations within the functioning of the international economy (and its institutional structure) as a necessary precondition for the plausibility of any alternative economic framework;
2. alternative ideas on national economic management that depart more radically from the neo-liberal model than the structuralist critique considered in chapter 7.

Those committed to the pursuit of more egalitarian forms of development have faced a particularly bleak political scenario over recent years – a reflection of such issues as the collapse of Eastern European 'socialism' and the supposed crisis of Marxist thought, the continuing intensification of global interconnectedness and the seemingly unassailable discursive dominance of neo-liberal ideas on development (see Brown 1996b). Such factors have spawned a series of wide-ranging debates over the prospects for the articulation of alternative strategies, central to which has been a stoical recognition of the constraints imposed by current global conditions.

In fact, the left has long been divided between those who have focused their political aspirations upon the pursuit of social transformation within individual countries and those who have maintained that any efforts towards progressive political change that were not international in their scope were doomed to failure, given the global reach of capital. Such conflicts resurfaced in academic debates within Marxism in the 1970s and 1980s and again, more recently, within discussions over the impacts of globalisation upon the possibilities for social transformation (Radice 1996). Within these recent exchanges, those most convinced that globalisation is occurring (and, moreover, that it represents a radical transformation in the dynamics of the international economy and a limiting of the possibilities for national economic management) often appear to argue that there is little that individual governments can do within such a context, implying that real alternatives can only arise from concerted political action at the global level. Indeed, even the more outspoken critics of the all-encompassing nature of the globalisation literature, such as Paul Hirst (Hirst 1997), recognise that the possibilities for the renewed national regulation and stabilisation of financial markets which they advocate are dependent upon concerted international action by major nation-states. The first part of this chapter, then, explores a range of possible transformations in the operation of the international economy as a prior imperative in formulating any realistic alternative to the neo-liberal focus on liberalisation, deregulation and export promotion.

The international dimension

For some observers, the debates between the IFIs and their critics reviewed in the preceding chapter fail to convince because of their lack of direct considera-tion of the external economic circumstances facing developing countries. They argue that discussions over the pros and cons of specific policy instruments within SAPs (or alternative frameworks for adjustment), whilst of obvious importance, must be seen as secondary to wider concerns about the inequalities of the international economic system. In this sense, SAPs are seen as just one, albeit extremely important, facet of a broader global economic system that needs reforming as a matter of significant urgency. Then, and only then, it is suggested, can discussions about the most appropriate form of national economic adjustment be embarked upon effectively. There is now a large litera-ture detailing this position which explores the ways in which the mechanisms of the international economy work to the detriment of the developing world (in terms of the organisation of international markets, the commodity terms of trade, the forms of operation of the major international institutions, etc.).

There are, perhaps, two basic positions within this literature. First, there is the argument that the international economy, as currently constituted, actively works to the detriment of developing countries' economic interests; a tendency which the free market-oriented reforms of recent years have certainly not addressed and, moreover, appear to have actively intensified. This position has, once again, focused attention onto issues which were central to the demands for a new international economic order which briefly flowered in the 1970s. Drawing upon certain strands of dependency analysis and the most pessimistic evaluations of the impacts of globalisation, attention has been drawn to such issues as the declining terms of trade facing the poorest commodity-producing countries (Singer 1991); the weak bargaining position of Southern countries within the World Trade Organisation (WTO) (and the detrimental impacts of its most recent settlements – Dasgupta 1998), the adverse impacts of the actions of, and the 'cut-throat' competition to attract, transnational investment (concerns over this were clearly articulated in the international opposition to the proposed Multilateral Agreement on Investment – see Davis and Bishop

1998/9) and the continuing burden of indebtedness afflicting many countries (Mihevc 1995: 55–84). Without addressing such fundamental issues, it is argued, whatever one's perspective on the theoretical merits of market liberalisation and the advocation of SAPs, in practice they will not lead to sustainable economic outcomes.

In essence, this perspective builds upon some of the concerns central to the structuralist critique of SAPs detailed in the previous chapter – the argument that there are structural causes (here considered to be of both national and international origin) underlying economic phenomena which SAPs, through their orientation towards internal policy issues, cannot address sufficiently. Such views are, however, combined here with a much more vehement critique of the negative impacts of adjustment and a much wider questioning of the goals of adjustment and the nature of development itself. Adjustment is, then, not only seen as ineffectual and misplaced, but also actively damaging in a much wider sense. From this perspective, the neo-liberal reform process is interpreted, not only as misguided in its understanding of individual economies, but also as pivotal in wider international transformations that have shifted the global economy towards a more 'savage' version of capitalism which has actively intensified the reproduction of global inequalities.

Second, there is a related, but somewhat distinct, argument that, whilst similarly focusing upon aspects of the global economy that impact negatively upon developing countries, has highlighted the ways in which such outcomes have been produced (or intensified) through a range of distorting practices that have been adopted by powerful Western economies. Such practices include: the restricting of imports in contentious economic sectors, such as those encompassed under the Multifibre Agreement (MFA) in the textiles industry; the replacement of traditional tariff-based barriers to trade with a range of non-tariff barriers (such as Voluntary Export Restraints (VERs) or anti-dumping legislation), that continue to restrict the access of developing country exports to key Western markets; and the continued endorsement of Western agricultural policies that subsidise agrarian exports, prohibit competition in internal markets and fundamentally distort international commodity markets (Woodward 1992: 156–161 and Dasgupta 1998: 151–162).

Whilst a wide range of commentators would agree upon the detrimental impact of these practices, there are sharply divergent views of their wider significance or quite how they should be responded to. One important point is that the continued adoption of such practices makes a mockery of the IFIs' demands for ever-greater liberalisation and deregulation of markets in the developing world – given the clear lack of commitment to such policies demonstrated by richer economies. As David Woodward (1992: 155–156) argues:

> the World Bank, through the dependence of most developing countries on its loans, is in a position to push them into adopting more market-oriented policies in their external trade; but, since it does not lend to the developed countries, it does not exercise any similar influence over them; and there is

no other effective source of pressure on them to adopt the policies they are instrumental in imposing on the developing countries.

At another level, however, the focus upon the distorting effects of government intervention is actually fairly reminiscent of neo-liberal positions on international trade – where similar arguments have underpinned their advocation of the continued liberalisation of international trade as the only effective way of combating Western protectionism and ensuring that developing countries are able to compete on a level playing field. Others, however, have a much less sanguine view of the potential for further rounds of trade liberalisation to secure any real changes in the global balance of power – either because of an *a priori* rejection of the mutual benefits of 'free' trade or because of the historical propensity for more powerful nations to circumvent measures designed to limit their unfair economic practices. This ability of the powerful to ignore the types of actions that are demanded of weaker states is termed 'brutal pragmatism' by Rosen (1997: 24) who, referring to the work of Theotonio Dos Santos, illustrates the idea in reference to the role of the United States within Latin America as follows:

> the US has been under pressure to reduce its own trade deficit. It has been pressuring Latin American countries to import US goods and to pay for their own trade deficits by bringing in 'hot money' with the high interest rates that strangle their own producers. And it has the power to do so. Neoliberal ideology has nothing to say here, argues Dos Santos. There is a rather 'brutal pragmatism' at work.

Both of these positions, therefore, albeit for somewhat different reasons, view SAPs with some suspicion and alarm because of their failure to recognise the importance of the negative impacts of the operations of the international economy upon developing countries (however those impacts are understood) or to suggest effective ways by which those impacts might be alleviated. These types of arguments have led to a range of different proposals (some relatively minor and others far-ranging and radical) for introducing fundamental changes in the way in which the international economy operates, as a necessary first step in the articulation of any alternative economic models at the national level.

Reforming the international financial system

Due to the key roles of the World Bank and the IMF in the emergence and promulgation of neo-liberal adjustment, they have, unsurprisingly, received significant critical attention from those campaigning for reforms in the operation of the international financial system (see, for example, the materials produced by the '50 Years is Enough Campaign', an international campaign organised around the fiftieth anniversary of the formation of the World Bank and its sister organisations – Danaher 1994). Obviously, given the range of different motivations for pursuing international reform, the types of transformations advocated have varied considerably. Some of the most polemical interventions have called for the complete abolition of the IFIs (or, perhaps, a gradual reduction in their funding – Rich 1994: 202) and their replacement with a range of smaller sectoral and regional organisations. Others have focused upon the organisational structure and forms of operation of the institutions as currently constituted. This has reflected an unease with how decisions are taken and policies formed within the IFIs – leading to proposals for the democratisation of their internal structures and voting procedures through, for example, changes in the system of voting rights for their major decision-making bodies (see the collection of essays in Griesgraber and Gunter 1996a). It has also been suggested that their activities might be made more accountable if their operations were to be considerably decentralised (Culpeper 1996, for example, explores the possibility of transforming the relationship between the World Bank and the Regional Development banks).

Somewhat wider in scope is the suggestion that the IFIs could be reconstituted so that they come under the direct control of the UN (or at least work in tandem with the UNDP) or that a new institution dedicated to facilitating debt reduction could be constituted (Woodward 1992: 121). Boutros Boutros Ghali, when he was Secretary General of the UN, called for the creation of a UN Economic and Social Council that would co-ordinate policy for the 'specialised agencies' (including the Bretton Woods Institutions), arguing that the 'present ambiguity and lack of co-ordination between the UN, the World Bank and the IMF can hardly continue for the next fifty years except to the detriment of the cause of development' (quoted in *Multilateral News* 2: 19: 5 October 1994). Whether the argument being made is for the internal transformation of the existing institutions, their replacement by new, more decentralised, organisations, or a limiting of their independence so that they come under the control of a new body with wider developmental aims and objectives, is, however, something of a moot point. Underlying each, there are a range of common sentiments.

First, there is the argument that, for some time now, the IFIs have not been fulfilling the role originally designed for them. It can be argued convincingly that the IFIs' role in the regulation of the global economy (eliminating large economic fluctuations; preventing huge surpluses/deficits and speculative capital movement; and recycling surpluses and preventing large exchange rate fluctuations) has been increasingly taken over by the periodic meetings of the G7 Group (adding weight to arguments over the lack of access of developing country governments to the real centres of economic power and decision-making). It is also suggested that the neo-liberal attitude of the IFIs in recent years has led to the abrogation of their original commitment to the prevention of speculative capital movement. Elson (1994: 521) points out that even relatively mainstream economists such as James Tobin have long been calling for measures that might slow down the speed of movement of capital through 'putting sand in the wheels' of the international money markets via transaction taxes or other similar measures designed to discourage speculation. Furthermore, at the same time as their traditional roles have been cut back, it is argued that the IFIs have been taking on new roles (involving much greater penetration into policy affairs which used to be seen as purely the preserve of elected governments) to which they are often clearly not suited.

Second, there is a shared conviction that the current regulatory framework of the international economy is not well suited to the rapidly globalising world of the 1990s, since it has not provided for an adequate response to the increasing ability of capital to escape from national levels of regulation. It is often pointed out that the current administration of the international financial system was obviously set up during a period of much lower levels of international integration. The massive transformations of recent years demand, it is argued, a radical overhaul of the institutional framework within which the global economy operates. There has also been sustained pressure, for example, for more effective regulation of the activities of transnational corporations – through the enhancement of existing corporate codes of conduct or the setting up of a more comprehensive regulatory framework (Brecher and Costello 1994: 123). Furthermore, the enhanced power bestowed upon the IFIs, following the economic crisis of the early 1980s, raises important questions regarding issues of national sovereignty and international accountability. It has not escaped the notice of many critics that, at the same time as the latest tranche of conditions imposed by the institutions has focused upon government accountability in the South, these same governments are being forced to transfer more areas of responsibility to remote unaccountable institutions in Washington (see Cahn 1996; Gillies 1996: 118–123 and Green 1995: 55).

Finally, there have also been calls for a more fundamental distinction to be drawn between aid (reflecting developmental goals) and loans designed to help tackle broad macro-economic imbalances. Under the SAP regime, this distinction has been lost and international aid, designed to meet basic human needs, has become diverted into crude economy-wide financial packages. The suggestion is that aid should recover its more direct developmental role through a form of conditionality that is much more closely tied to individual governments' record on meeting the basic needs of their populations rather than broad macro-economic criteria. Such ideas have been associated with proposals for the drawing up of a World Social Charter defining universal social rights, together with related mechanisms for their enforcement (see also the section on social clauses below) – although other critics have questioned what they have interpreted as a further enhancement in the power of unelected international institutions (Waterman 1996: 172).

Issues surrounding international debt

Intimately linked to the calls for the transformation of the international financial system considered above, has been a campaigning focus upon debt cancellation. Here, the basic argument is that whatever the pros and cons of SAPs, the fundamental problem facing developing countries is the continued necessity of meeting interest payments on their national debt – a task that years of austere adjustment have not managed to ease for many countries (despite it being the basic rationale for adjustment processes in the first place). Addressing this 'day-to-day' drain of resources is seen as a much more urgent task than critiquing the failures of the adjustment programmes designed to maintain debtor states' ability to service their debts. In some campaigning quarters, such positions have been reflected in demands for the total 'forgiveness' of the debt or, at least, the debt of the poorest countries (see, for example, the global efforts of the Jubilee 2000 Campaign – http://www.jubilee2000uk.org/) or, perhaps, some form of global income redistribution through increases in overall aid levels (e.g. Frankman 1992). Others, however, have been quick to point out that, whilst an end to debt repayments would certainly have immediate beneficial impacts, it would not address the reasons underlying the growth of the problem in the first place and would leave the inadequate systems for dealing with such issues in place, so that the problems would no doubt recur in the future. Elsewhere, therefore, there have been demands for a much more co-ordinated approach towards the whole issue.

There have recently been a series of major attempts to force 'debt' back on to international agendas, such as the Latin American parliament's (PARLATINO) recent challenge to the legality of continued debt servicing at the International Court of Justice (Rosen 1997). This stress upon the need for a more co-ordinated approach to the whole issue of debt and its management effectively leads us back to the more generalised arguments regarding the transformation of the international financial system or, at the very least, for changes in the way in which the current system has responded to the continuing financial crisis of many countries. Much has been made of the inadequacies of the case-by-case approach that the institutions have adopted and many analyses have called for much more effective international co-ordination that recognises the severity of the problems being experienced much more seriously and sympathetically.

Other institutional avenues

The reform of the international financial organisations and a more effective treatment of developing world debt represent major goals for many critics; although it is also recognised that significant steps in such directions will not happen rapidly, if at all, given current political realities. Significant attention has also, therefore, been devoted to the search for other international arenas and initiatives where alternative viewpoints might be articulated, or where pressures for reform of the institutions might be effected. As we have seen in earlier chapters, many other international organisations have been far from convinced by SAPs and the neo-liberal stance of the IFIs. We have already discussed attempts made by the regional United Nations institutions to suggest alternatives, as well as the critique of the social impacts of adjustment articulated by UNICEF.

Perhaps the most obvious arena for the development of critiques and the evolution of alternative frameworks (given the impacts of SAPs upon the relations between capital and labour and overall working conditions across the globe), however, would appear to be the International Labour Organisation (ILO).

The primary objective of the ILO, as stated in its constitution, is to ensure the protection of fundamental workers' rights. Given this, it would appear to be the natural institutional arena within which the interests of labour might be defended against the global attacks on employment, wages and working conditions which neo-liberal adjustment has promulgated. Nevertheless, there is scant evidence of this potential being realised. This is not that surprising, given that the ILO has never been an organisation which represents labour as such. It is, rather, a tripartite organisation that attempts to accommodate the interests of governments, employers and workers (at each annual conference, for example, the individual member nations are represented by two government representatives and one each from the national trade union movement and the employers' organisations). Nevertheless, in previous decades, the ILO's focus upon labour issues had made it a natural 'lead player' in the evolution of ideas on development (during the 1970s, for example, the ILO had been one of the major architects of the emergence of a basic needs agenda amongst the IFIs – see Kitching 1982). Its relative silence on the social impacts of adjustment and the formulation of more labour-friendly alternatives is, therefore, even more surprising.

Despite this, some observers still see the ILO's focus upon building international support for the protection of basic labour rights as a possible counter-weight to the over-economism of the IFIs. The ILO attempts to achieve this through the establishment and ratification of international conventions on labour issues. By the early 1990s, there had been 174 international conventions and 180 recommendations concerning basic human and workers' rights, often referred to as the International Labour Code. Whilst of obvious significance, there have been two major limitations to this process. First, the fact that each government has to be persuaded to ratify any of the conventions passed and, second, the fact that the ILO has no real power to penalise nations who fail to comply or break conventions which they have ratified. The willingness of governments to ratify conventions, in fact, varies widely; a point which has led to some criticism of the ILO's preoccupation with the whole system of convention building. In 1983, of the 159 conventions existing at the time, Spain had ratified 90, the UK and France had both ratified 80, the USSR 57 and the USA had ratified only 7 conventions (this had increased to 11 by 1994). Even where a country has ratified a convention it may only express an intention to act and not a firm commitment (Bendiner 1987; Richardson 1983; Brecher and Costello 1994 and Galenson 1989). Overall, the ILO remains an important potential source of alternative ideas and checks on the extremes of neo-liberal reform but, to date, has generally appeared strangely muted in the face of the growing problems faced by labour movements internationally.

Social clauses

Recent years have seen growing international support for a much more specific attempt to highlight abuses of basic human rights and unacceptable working conditions (and, on occasion, environmental destruction) – the advocation of social clauses within international trade agreements (Shaw nd). These proposed clauses, often based around criteria derived from the International Labour Code, consist of additions to international trade agreements which would facilitate the application of sanctions (involving banning or restricting exports – or revoking any preferential trading status) against countries that fail to observe certain minimum labour standards.

Not surprisingly, such proposals have inspired heated debate. It is worth noting, however, that the whole question of social clauses is one of the few elements of progressive reformist agendas that has received attention at the highest international levels. The WTO, for example, has debated the issue quite extensively, although it shied away from any commitment to setting up an exploratory working group. As would be expected, most of the major financial organisations are against the idea – viewing social clauses as representing a slide into protectionism. The issue has also been raised at the ILO on several occasions (1973, 1988 and 1994) but the case for social clauses has failed to gain 'across the board' support – with union representatives largely in favour, employers against and governments divided. Some of the most detailed proposals for the implementation of social clauses have come from the International Confederation of Free Trade Unions who, obviously, support their implementation; although interestingly several regional trade union organisations have come out strongly against the idea (as have several Third World governments – it was the Latin American negotiators, for example, who 'scuttled' proposals to set up a working group in the WTO in 1994) (Shaw nd: 22–26; Brecher and Costello 1994: 135).

The arguments in favour of social clauses are obvious – the moral and political cases for the protection of basic human rights and restoring a concern for social justice to international institutional debates. In addition, low labour standards can be seen as imposing substantial longer-term economic costs through their impacts upon labour productivity, social unrest, etc. More widely, the advocation of social clauses can be seen as part of an attempt to establish a freer market in labour to match the free markets rapidly being established in trade, finance and industry. If workers cannot migrate to the West and, therefore, earn the international market price for their labour, the argument runs, how can they be denied access to a mechanism that might improve that price in the countries where they have to stay?

Given the weight of these arguments, why has there not been more universal acceptance of the case for social clauses amongst international NGOs and developing country governments? One interpretation might be to stress the ways in which repressive Third World governments might seek to perpetuate the over-exploitation of their workforce, but this does not explain the opposition of some trade union organisations. A good part of the unease with social clauses expressed in the developing world reflects a concern that the international advocation of such clauses, rather than offering a means of protecting the working conditions and living standards of the poorest, largely reflects the interests of protectionist forces in the West who see social clauses as a potential mechanism for combating some facets of Southern competition in Western markets (a possibility which, the cynical might suggest, helps to explain the seriousness with which the case for social clauses has been explored at the highest levels). The unease amongst Southern governments and organisations, therefore, reflects the fear that social clauses are merely a thinly disguised tool of Western protectionism or, perhaps more appropriately, too open to manipulation towards that end. Some evidence from countries that have unilaterally adopted similar measures suggests that social clauses could be susceptible to very obvious political manipulation. The Reagan administration in the United States, for example, used the US Trade Act of 1988 in some very suspect political manoeuvres (Brecher and Costello 1994: 132).

More generally, the advisability of seeking to pursue a renewed process of progressive regulation of economic activity within the arena of the management of international trade – a notoriously conservative set of institutions – has been seriously questioned. It would, perhaps, make more sense to try and incorporate any commitments to the greater protection of labour rights into an international effort to overhaul the ILO and increase the commitment of Western governments to its principles and conventions. There is also a feeling that the pursuit of social clauses has tended to isolate the issue of labour conditions from broader issues of international economic inequality – leading to questions surrounding the ability of inter-*national* agreements of this type to deal with the global activities of transnational corporations and the limitations of a restricted focus upon those actually engaged in wage labour.

The limitations of international reform

The preceding discussion has illustrated a range of positions on the unequal nature of the functioning of the global economy. Obviously, this brief survey has not been exhaustive, but it has illustrated the breadth of the case for international reform. One further factor that should be addressed, however, is the way in which apparent campaigning successes (in other words the achievement of policy, or even institutional, changes) do not always produce the results intended because of the abilities of the institutions themselves and other dominant groups to divert, weaken or co-opt criticism (as was outlined in our discussion of the Bank's reaction to specific policy critiques in the previous chapter). The point being made here is not that campaigning at the international level is pointless (it obviously has brought modifications to specific policies and raised the profile of issues that would otherwise not have figured in global debates) but, rather, that it cannot be seen as a panacea and must be complemented with approaches that are more fundamentally concerned with what happens on the ground in individual countries. As Duncan Green puts it:

> Although the technological genie which drives economic globalization cannot be put back in its bottle, changes in the management of the global economy could ease the task of switching to a growth-with-equity development model. ... But Latin America needs much more than a new deal with outside forces, however important they are.
>
> (Green 1995: 207–208)

Pleas for fundamental changes in the functioning and dynamics of the global economic system and its institutions must, therefore, be tempered by political realism. The simple fact is that the political will to achieve such reforms is very unlikely to exist amongst the global powers and, if that is the case, it raises serious questions about concentrating upon international reform as the panacea for 'developmental' problems. Even if such transformations were to occur, however, they could only ever be part of the story because, if the existing national structures of production and exchange were to continue in their present form, the vast majority of people would gain little. As argued by Frances Stewart:

> Governments may be relatively powerless to steer the course of world markets, but they can protect their own people from the excesses of the market. Conversely, there could be a good international system which protects countries, but if gross inequalities emerge within countries, such a system will not succeed in protecting poor people – and since it is people rather than countries with which we should be concerned, domestic measures to control the market and protect the vulnerable are generally more important than international. This is, indeed, just as well, since while it is quite easy to identify changes in the international system which would help prevent international inequalities, it is much more difficult to envisage (and even more to bring about) the political conditions in which such changes are plausible.
>
> (Stewart 1995: 164–165)

Stewart's concern for domestic measures to control the market and protect the vulnerable in the context of limited international room for manoeuvre is important in that, whilst everything should be done to pursue reform at the international level, it is at the national and local levels that real alternatives to neo-liberalism must be found, although such alternatives will, no doubt, find themselves restricted by the international considerations that we have touched upon thus far (bringing us back once again to the controversies over the extent and impact of globalisation with which we started this chapter).

Restructuring the socialist alternative

Moving to the national arena once again, the most immediate problem encountered by those opposed to neo-liberal adjustment relates to the apparent collapse of the major policy alternatives to that model. For most of this century, there have been two major alternatives to free market capitalism (of which neo-liberalism is the latest incarnation). First, what we might loosely term 'reformist' versions of capitalism (the breadth of this term can encompass the interventionist Keynesian approach to macro-economic management, the social democratic politics with which it was often associated and the various terms used to suggest the interventionist role of the state within such models – statist or dirigiste strategies) which sought to control capitalism in the pursuit of specific socio-economic aims and objectives. Second, various radical socialist alternatives that proposed a much more fundamental break with capitalism.

Many of the criticisms of SAPs and the alternatives suggested in chapter 7 owe much to reformulations of reformist approaches to capitalism, few of which suggest a decisive break with the dominant neo-liberal development discourse. In fact, most reconfigurations of such positions today reflect an accommodation with elements of the dominant neo-liberal perspective. As many neo-liberal theorists have come to recognise the limited applicability of the more extreme models of adjustment to market forces, and more reformist minded economists have conceded aspects of the neo-liberal critique of the state and faced the cold reality of increasingly liberalised and global markets, so the differences between reformist and neo-liberal approaches to the South have become increasingly blurred. It was, of course, the idea of socialism that stood as the major alternative to Western capitalism throughout most of this century through its profound critique of the nature and dynamics of the latter and the advocation of a radically divergent egalitarian alternative. Today, however, following the collapse of the Soviet bloc and the end of the Cold War, it has become far more difficult to establish a firm and definitive divide between capitalist (be it orthodox or reformist) and radical or socialist approaches towards development. Many avowedly socialist political parties appear to differ little in their policy recommendations from their social democratic, or even in some cases neo-liberal, counterparts.

Few treatments of the adjustment process (or even of the alternatives to it) pay much attention to specifically socialist alternatives. Generally, existing socialist regimes are interpreted as anachronisms, living on borrowed time and awaiting the unleashing of pro-democracy forces which will inevitably see them pass into history in the same way as the Eastern European socialist states. It cannot be doubted that the world facing socialist movements has changed considerably since the Cuban revolution at the end of the fifties or, indeed, since the emergence of Marxist regimes in Angola and Mozambique or the Sandinistas in Nicaragua during the late 1970s. Whatever one's thoughts on the ex-socialist regimes of Eastern Europe, and however far individual socialist governments or movements in the Third World attempted to distance themselves from the more unsavoury aspects of such regimes, the enormous changes in global geopolitics caused by their collapse and disappearance have made any attempt to pursue 'socialist' policies in the years which have followed much more difficult. An important alternative source of financial resources obviously dried up and, even worse, was accompanied by descent into international depression and the emergence of an international financial regime that granted precious little leeway to the pursuit of alternative economic programmes. Similarly, the events of recent years have seen 'socialist' become a word associated with the past, reflecting economic failure and the abuse of basic individual human rights and certainly not as a potential source of realistic alternative development policies. As Cavarozzi (1993: 152–153) suggests in relation to the Latin American left specifically:

> The ideological crisis of the Latin American left has certainly intensified with the collapse of the communist regimes of Eastern Europe and the Soviet Union. ... Their mere survival was enough to fuel illusions of replicating them in Latin America, purging them of their negative traits – that is, bureaucratic control and political totalitarianism – and preserving their relative egalitarianism. In this sense, the evaporation of the Communist regimes in Europe – and the confirmation of the fact that almost the entire populations of those countries perceived those systems as oppressive and stagnant – has affected the credibility of the entire Latin American left. ... There are still some activists and ideologues who argue that there are still a thousand million Chinese living under a communist regime and that there is a remnant of virtue in the Vietnamese and Cuban regimes. But this can only be understood as the desperate search for a psychological refuge from the wreckage of the political and economic parameters that had prevailed for seven decades.

There is certainly a good deal of truth in Cavarozzi's position but it would be a mistake to assign all of the experiences and, indeed, achievements of socialist governments and movements to the wastebasket of human history. Whilst many expressly socialist regimes have disappeared over recent years under the weight of their own contradictions (or the pressure of unfavourable international events), a number of regimes have persevered (if not prospered) into the final stages of the twentieth century. More importantly, socialist political parties and movements (of many different orientations) remain key political actors in many countries (in Latin America alone 30 per cent of the electorate voted for left-wing alternatives in the mid-1990s – NACLA 1997: 5). Much of the most vociferous criticism of the social impacts of adjustment has come from these organisations; their major challenge remains quite how to transform that opposition into a workable progressive agenda for the economy. The search for such a programme hinges around the need (1) to go beyond the state-centred approach to the economy which, to differing degrees, had dominated socialist regimes in the past, (2) to ensure that any alternative is not simply a form of 'managing' neo-liberalism (in other words the reformist attempt to ameliorate the social costs of neo-liberal policies without offering a fundamental alternative) and (3) to re-evaluate the relationship between socialism, democracy, popular participation and the development of civil society. This section considers some of the debates that have occurred amongst the left as it has grappled with these three elements.

The state-led model

In considering the possibilities for the articulation of a workable socialist alternative, the obvious place to begin is by looking at the developmental performance of existing socialist regimes. Twenty years ago, there would have been a wide range of material to consider that detailed such experiments, as there were a broad range of socialist regimes in existence (Utting 1992). Today, many of those regimes have fallen from power and those that remain have undergone such transformations that many are hardly recognisable as 'socialist' at all. Nevertheless, it is still worthwhile revisiting some of those experiences.

The classical writings on the construction of socialism were largely based around economic strategies that assumed the existence of an economy that had already experienced a high degree of productive development. For much of the earlier decades of this century, this produced a dominant Marxist view that any transition towards socialism was impossible without the prior transformation of society under the expansionary influence of capitalism (see Warren 1980 for a reassertion of this view). In Latin America, for example, the 'official' communist parties of the region generally subscribed to such analyses and, accordingly, they (and other leftist forces within the region) tended to prioritise political alliances with the more nationalist-orientated sectors of the emerging bourgeoisie, rather than revolutionary mobilisation towards a more immediate attempt at constructing socialism. This tendency was to change markedly following the Cuban revolution and the subsequent emergence of dependency theory within Latin America (see Kay 1989 for what is probably the most accomplished survey of this literature) as well as the growing association of socialist ideas with the nationalist revolutions in Africa. Under the influence of these emerging theoretical positions (and the examples of praxis in Cuba and China, etc.) a body of literature on socialist transition, rooted in the concrete experiences of developing countries, gradually emerged – although the Soviet influence was to remain considerable (as a non-capitalist route to socialism was gradually adapted to the orthodox communist canon – see Solodovnikov and Bogoslovsky 1975).

In the years that followed, a great number of regimes came to embrace some sort of variant of the socialist approach and, as Simon (1995: 709) notes, 'almost as many forms arose as the number of countries claiming such a pedigree'. These ranged from the limited number of expressly Marxist regimes, 'through various hybrids to the large group of essentially populist, nationalist governments cloaking their ideology in socialist rhetoric' (ibid.). Nonetheless, despite these different individual circumstances and orientations, what united the various socialist regimes that emerged in the South was a shared emphasis upon the role of the state (in its productive and administrative roles) in guiding society towards a (supposedly) more egalitarian form of development.

The key dilemma facing socialist regimes in their search for the most appropriate forms of state intervention in the pursuit of their objectives was the delicate trade-off between prioritising overall output levels and developing the forces of production (hence hopefully ensuring the ability to finance its social programmes) or attempting to, more rapidly and radically, transform the relations of production (through, for example, land reforms designed to break the power of traditional agrarian elites). The latter alternative may itself involve substantial financial costs due to inflationary impacts upon labour costs or the loss of the economic experience of business leaders in important sectors of the economy. In practice, financial concerns (often coupled with a mistrust of the peasantry and smaller-scale production units) tended to lead towards policies that prioritised overall gains in productivity, with changes in social relations relegated to those that would gradually be produced through the expansion of the dominant state productive sector (Griffin and Gurley 1985 and Fagen *et al.* 1986).

This approach, often termed 'state-centred accumulation', owed much to the legacy of Eastern European thinking, and led to emphases upon the rapid promotion of industrial activity (largely financed via the extraction of surpluses from the agricultural sector), the subsidisation of basic necessities, the collectivisation and rapid modernisation of agriculture, and centralised comprehensive planning systems (White 1983: 1; Brown 1996a: 277 and Harris 1992: 87). By and large, the record of this state-led model of socialism, even though it was centred around the rapid development of productive forces, was not generally successful in economic terms. Whilst there were certainly some successes in terms of the development of social provision and the, at least partial, eradication of inequality, these advances were generally jeopardised by a lack of economic success. As expressed by Forrest Colburn (1994: 63):

> an atmosphere of crisis and austerity has become the normal condition of economic policy in the revolutionary setting. Rosy expectations are frustrated. While some states, notably Cuba, reduced inequality, economic growth has in most cases been disappointing, and in some cases absent altogether.

The sorts of problems identified include: the poor productive record of state enterprises, the limited capacity of the state to effectively carry out the expanded range of tasks that have been expected of it, poor levels of productivity given the lack of individual motivation afforded within collectivised economic activities, the centrality of the export sector (given its vulnerability to external circumstances and destabilisation and the distributional impacts of its prioritisation) and the contradictions between production, consumption and distribution (Harris 1992; Colburn 1994; Vilas 1990). There is also, of course, the issue of the lack of basic democratic rights and the over-arching role assigned to the vanguard party within the traditional state-centred model and the political costs that this has often involved (see Post and Wright 1989 for example). There is a massive literature on the problems and pitfalls of the state-led model of socialism and it is certainly not worth revisiting those arguments in detail here – suffice it to say that, with some exceptions, few would now actively defend the record of such strategies. The stress on productive development under the auspices of strong state control leads Carlos Vilas (1992: 16) to conclude that 'much of what is conventionally considered socialist economic policy in the underdeveloped world is simply a leftist version of desarrollismo, the notion that technological advance equals development'.

Most assessments of the developmental record of the state-led strategy of socialism in the South would, then, encompass the sorts of damning assessments considered briefly above. Before moving on to consider some of the alternative directions in which the attempt to rethink socialist economic strategy has led, however, it is worth applying a number of caveats to the discussion thus far.

First, it is easy to overstate the economic failings of the state-led socialist strategy and certain social achievements under that model need to be recognised. Cuba, for example, despite the problems its dependency upon the Soviet bloc has bestowed upon it and the clear limitations on individual freedoms, certainly boasted health and education systems that, at least until the economic crisis of recent years, were unequalled anywhere in Latin America, as well as a level of equality that was unprecedented within the region. Even the poor economic record ascribed to the model needs some qualification. For the period 1960–74, for example, Jameson and Wilber (1981) found that a loosely defined group of socialist countries actually grew at a faster rate than non-socialist countries.

Second, many negative reviews of the economic performance of socialist states tend to ignore the external aggression which many of them have had to contend with (such as the US blockade of Cuba and the covert war fought against revolutionary Nicaragua or the South African destabilisation of Angola and Mozambique) although the special treatment which they received from the Soviet bloc should equally be borne in mind.

Third, it is easy to over-generalise about 'socialist' regimes (see White 1983: 10). There are different lessons to be drawn from the experiences of those countries most closely linked to the Soviet bloc when compared to those of more open and pragmatic regimes such as the Sandinistas in Nicaragua or specifically African variants of socialism such as the Ujamaa experiment in Tanzania. Whilst each of these examples shares many of the characteristics of the state-led model described above, it is worth stressing both the individual priorities of those regimes and the fact that there have always been socialist traditions that differed substantively from that general tendency (even if they have not been terribly influential in the trajectories of socialist-inspired movements and regimes in the Third World).

Fourth, whilst socialism has been written off as an irrelevance in many quarters it continues to inspire many people and movements; indeed political parties retaining a commitment to such ideals remain, as has already been pointed out, important political actors in many regions. Nevertheless, given the limitations of previous socialist experiences in government and the problems and contradictions experienced by those that still exist, quite what such movements stand for and what types of economic programmes they might administer, were they to achieve political power, remains very much open to question and the subject of much internal deliberation. The nature of some of those areas of debate and the question as to whether they might lead to the articulation of clear economic alternatives, substantively different from the dominant neo-liberal model (and the state-centred reformist and socialist models of the past), forms the focus of the remaining parts of this chapter.

The state of the left today

A hankering for the past

Some orthodox socialist parties have continued to espouse traditional state-led economic strategies revolving around the nationalisation of the most important sectors of the economy, far-reaching processes of collectivisation and the state administration of markets. The interesting thing to note is that the advocates of such strategies appear not to have noticed how even those regimes which often form the focus of their aspirations have undergone considerable transformations in recent years. Even Fidel Castro, for example, has suggested that any leftist governments achieving political power under contemporary circumstances would not be advised to adopt sweeping processes of nationalisation.

Another traditional tendency within some socialist strategies in the Third World is that, given the exploitative nature of international economic relations, engagement in international markets is something to be avoided as far as possible. Particularly associated with certain elements of the dependency school, such autarchic strategies have not often been attempted and, where they have been, they have not produced particularly impressive economic results and have often been associated with savage repression of the dissent that they have engendered. Given the contemporary processes of globalisation and the increasing size of urban populations, autarchic experiments (by necessity largely agrarian in orientation) appear neither politically feasible nor economically desirable in the current context. Despite its limitations, this type of perspective is also still articulated on the contemporary political stage. It is most strongly present within positions held by the most violent sections of the left, through, for example, what Cavarozzi (1993: 155) refers to as the idea of 'Millenarian violence' which centres around the need to make a dramatic break with the past through the forced re-education of the urban sectors or the 'profit-maximising' peasantry (Sendero Luminoso in Peru would appear to be representative of these ideas). Common to both of these positions is a continued adherence to the Leninist model of socialist politics where socialism is pursued through the autocratic actions of a verticalist vanguard party that monopolises political power (although it should also be recognised that a more localised form of autarchy is also advocated by 'deep green' environmentalists).

Embracing the neo-liberal model

> In some instances, revolutionary elites were forced out of power. More commonly, revolutionary ideals were just abandoned by their erstwhile proponents.
>
> (Colburn 1994: 89)

> Self-styled revolutionaries, with rare exceptions, have yet to get beyond the state-centred vision of socialism so discredited by the experiences of the Soviet Union and Latin American populism. The parties of the Left are in many ways as conservative and authoritarian as those of the right. ... During the transition to electoral regimes, most revolutionaries either stuck to their guns, decrying democratization as false, or embraced the establishment, abandoning altogether their hopes of achieving socialist reform.
>
> (NACLA Report on the Americas 1993: 12)

Whilst some socialist movements continue to cling to the economic models of the past, others appear to have abandoned their socialist ideology in all but name (and sometimes that as well!). Thus, some traditional communist parties have renounced their previous orientation and embraced the market even more obviously than some social democratic parties – reflecting either a renouncement of anything approaching socialist ideology or a reversal to the 'socialism can only emerge from the eventual contradictions of a fully developed capitalism' position of the past. The Chinese opening up to market forces (as well as that of some of the communist parties of Eastern Europe and several 'socialist' regimes during their death throes in Africa) and, in particular, the decisive new orientation towards serving export markets, can also be interpreted in this light. Such strategies have, despite an often-repeated claim to be building 'a socialist market economy' (Miliband 1996: 15), left little indication of what might demonstrate their continued 'socialist' orientation, save the maintenance of the traditional authoritarian one-party political system.

In a sense, those societies living under such regimes have the worst of all worlds – the social impacts of export-orientated capitalism and the authoritarian political system of one-party socialism. Before we totally write off the Chinese system for its obvious political brutality, environmental disasters and neo-liberal-inspired restructuring, it is worth sounding a note of caution. Whilst the working conditions and lack of worker representation in the burgeoning export enterprises in the industrial sectors have been well documented, the transformations that have occurred in rural areas over the last few decades are less well known. Here, the opening up of rural markets and the setting up of state-funded rural township enterprises (amongst other policies) have seen rural incomes practically quadruple. There has been a mixture of state protection and encouragement of small-scale agricultural and industrial production with market incentives. However, in contrast to the neo-liberal approach towards market reform in rural areas, land has remained in the hands of the state, avoiding the escalation in rents and speculation in property that has occurred elsewhere (Burbach *et al.* 1997: 92–93).

Nevertheless, there are a number of factors that make it debatable whether the Chinese experience is likely to prove attractive to other socialist strategists. First, whilst the strong Chinese state has enabled it to avoid the political chaos that accompanied market-led reform in the former Soviet Union (Marshall 1998: 286), it also remains an autocratic, profoundly anti-participatory presence that has effectively strangled any political reactions to declining living standards and increasing inequality and failed to offer any space for the articulation of ideas and strategies reflecting real grassroots interests. Second, whilst rural development schemes may have proven successful, the dominant force within the Chinese economy is undoubtedly the highly divisive, income concentrating, industrial activity of the export sector (Cannon 1994: 523) and it is far from clear how, or indeed if, the inherent social contradictions will be dealt with. Finally, the sheer size of the Chinese economy (and perhaps most importantly its potential domestic market) has enabled it to enjoy a very different relationship with the international community to that of other smaller developing economies who are much less able to 'pick and choose' which elements of the neo-liberal agenda they wish to adopt.

The search for renewal

Moving away from the experiences of existing socialist regimes and the most orthodox political parties, other leftist movements have been caught up in an urgent search for renewal – for a new strategy (or range of strategies) that might prove capable of offering a workable alternative to the state-centric models of the past, without a wholesale capitulation to neo-liberal deregulation.

Shouldering the social costs of neo-liberalism?

In its more reformist guises, the left has played an active role within the structuralist attempts to articulate a more interventionist response to neo-liberalism which we considered in the preceding chapter. This type of approach, predicated upon extremely pessimistic readings of global economic circumstances (or a genuine acceptance of major parts of the neo-liberal critique of the state), represents, in essence, an effort to manage the social impacts of neo-liberal reform through a more progressive distribution of its costs and benefits – a more socially responsible model of neo-liberalism as it were (often conceptualised under the idea of the social market).

This approach is probably most clearly developed in Jorge Castañeda's (1993) exhaustive review of the circumstances facing the Latin American left – *Utopia Unarmed*. In essence, he calls for the left to abandon the search for a socialist utopia and instead engage in the struggle being undertaken by organisations such as CEPAL to accept global economic circumstances as they are (and some of the insight of the neo-liberal critique of previous development models) but to look for a way of combining this with a more sustainable and equitable approach, perhaps based on the legacy of the West European welfare state and the Asian approach to state–market relations. His argument is that it is only through 'formally and sincerely accept(ing) the logic of the market' that the left will attain the credibility to challenge the most starkly anti-interventionist and brutal forms of market capitalism (Castañeda 1993: 432). The specific policy measures proposed are familiar and include regional integration and a form of state activity that involves 'protection without full-scale protectionism, regulation without stifling the market, state ownership without operating a command economy and competition without savage capitalism' (Castañeda 1993: 441). Central to any of this, he argues, is the effective reform of the taxation system followed by a more effective tackling of poverty (Green 1995: 191 and Portantiero 1993: 17–20). The political challenge, according to Castañeda, is to convince the business leaders of the region (and the US) that the social costs of neo-liberal economics have so heightened the gap between rich and poor that it is leading towards an inevitable social explosion unless ameliorative action is prioritised (Castañeda 1993: 428).

Clearly, Castañeda's approach is not very different from the structuralist strategies discussed in the previous chapter and it shares their somewhat techno-cratic nature, in that it revolves around an economic strategy that relies upon essentially technical and institutional interventions in the market (Cole 1996) and a political strategy geared towards convincing sections of the elite that a socially mediated version of neo-liberalism is in their best interests. It is this political dimension – the somewhat naive suggestion that somehow Latin America's elites will respond to the threat of social violence in the manner hoped for – which is probably the weakest element of Castañeda's proposals (which is surprising given the depth of political analysis elsewhere in his book, Hammond 1995: 119). Finally, despite his protestations to the contrary, Castañeda's position in reality amounts to little more than an appeal for European-style social democracy and this is somewhat ironic given the current state of the European left which appears to have gained a certain political initiative in recent years through moving closer to the neo-liberal model and abandoning the more interventionist type of approach that Castañeda is looking to embrace. As Miliband (1996: 17) argues:

> the leadership of Western social democratic parties in France, Germany, Britain, Spain and even Sweden have moved fairly sharply to the right over the years and have come to accept without much difficulty the market economy and privatization. They qualify this acceptance in various ways, but they do not provide anything like a serious counterweight, in ideological, political, programmatic, or practical terms, to conservative or neo-liberal parties.

This type of strategy has been fairly typical of the party political left over recent years in Latin America and appears to be a throwback to the 'popular front' political strategies of the early decades of this century. As Ellner (1993: 7) suggests, however, there was, in that earlier period, a much clearer political strategy (that of supporting nationalistic fractions of local capital against feudal elites and US interventionism) which appears to be lacking in the current electoral strategies.

A more radical agenda?

In a sense this brings us back to the problem identified at the end of the last chapter – a political agenda dominated by underlying neo-liberal ideas that appears to preclude the evolution of any more radical alternatives. The discrediting of the state-led strategies of the past, be they capitalist or socialist, has left those committed to more egalitarian development strategies struggling to articulate an alternative that does not repeat the mistakes of the past and yet also provides something more than a choice between different variants of neo-liberal capitalism (Burbach *et al.* 1997). As William Robinson (1998/9: 125) puts it, these circumstances have created a situation where:

> many leftist parties, even when they sustain an anti-neoliberal discourse, have, in their practice, abdicated earlier programmes of fundamental structural change in the social order itself. Their programmes in the 1990s were confined to strategies of state intervention in the sphere of circulation to achieve limited internal redistribution, while respecting the prevailing structure of property and wealth and the model of 'free market' integration into the global economy.

As such, there is a serious danger that, whilst the extreme social polarisation produced through neo-liberal adjustment might render socialist alternatives more attractive to disillusioned electorates (and that is by no means certain – see discussions on this in chapter 5), the lack of a viable economic strategy for achieving socialist political goals will mean that leftist governments may well end up 'managing neo-liberalism' by default – providing a slightly enhanced social safety net and yet doing little to challenge the prevailing patterns of income and resource distribution or the dominant attitudes towards society and politics. In the light of such arguments, many on the left are suspicious of reformist efforts such as Castañeda's, seeing them as little more than a capitulation to the contemporary dominance of neo-liberal ideas and an abandonment of any pretence at more radical social transformation. For those with such concerns, there is an urgent need to articulate a more expressly socialist alternative which, nonetheless, remains frustratingly elusive.

In Latin America, one of the arenas for discussing the formulation of just such an alternative has been provided through the establishment of the São Paulo Forum launched by the Brazilian PRT in 1990. The 1992 meeting of the Forum in Managua saw representatives from over fifty organisations from seventeen different countries attempt to outline the basis for an economic alternative to neo-liberalism (this number had increased to 112 member parties by the time of the 1996 Forum in San Salvador – NACLA 1997: 5). The declaration produced from that meeting started from the premise that neo-liberalism could not be amended since it is part of an unjust economic order, echoing our discussion of the wider calls for international economic change outlined above, rather than Castañeda's accommodation with the neo-liberal rules of the game. None the less, quite what the alternative might be did not progress far beyond vague commitments to autonomous economic development, meeting the basic needs of the majority, providing a fairer distribution of wealth and property, enabling a leadership role for popular grassroots organisations and a central regulatory role for the state (Green 1995: 195–196).

At this juncture, it is worth briefly returning to a number of facets of our discussion of the experiences of socialist regimes in power – whatever our view of the democratic credentials of such regimes their attempts to reform the state-socialist models of the past, without a wholesale adoption of neo-liberal reform, may prove instructive. Peter Utting (1992) provides an excellent discussion of the attempts made by a range of socialist regimes during the 1980s to switch from state-dominated economic strategies to some new model which, whilst less centralised, still retained a clear socialist orientation. As Utting points out, a number of such regimes had realised the limitations of the state-led model well before the generalised crisis of the late 1980s and the specific circumstances of that period were certainly not the only factors pushing them in new directions (see also Fitzgerald and Wuyts 1988). The productive limitations of state-run enterprises and the incapacity of the state to fulfil the role expected of it, together with a recognition of the complexities of social identity and the declining size of the social sectors comprising their traditional political constituency, had already begun to convince certain sectors of the need for a fundamental re-evaluation of how socialism was conceived.

In essence, this re-evaluation stemmed from the need to confront two inter-related problems centring around the role of the state. First, a growing accumulation crisis, whereby the state-led model, which, as we have seen above, had never been exactly over-successful productively, was increasingly unable to meet even the basic consumption or investment needs of society due to its increasingly apparent inefficiencies in the context of reduced external co-operation and international recession. Second, a crisis of hegemony, whereby regimes found themselves increasingly unable to maintain control over the direction of social change and economic development and forced to recognise the complex fragmentary nature of class and other social relations within their societies and their political impacts (Utting 1992: 256–258). As David Simon (1995: 713) argues, the changing circumstances allowed 'previously suppressed domestic resistance to authoritarian rule and opposition, deriving from the inability of these regimes to deliver material needs and social welfare to their people, ... to resurface to challenge regimes of all ideological hues with greater vigour'.

These crises produced a number of different responses. Some regimes sought, as far as possible, to maintain the dominant role of the state within the economy, although most, as they desperately sought for strategies that would improve their economic performance, at least flirted with the introduction of markets in some areas. The types of strategies embarked upon included measures that recognised the continued importance of the peasantry to the agricultural sector and sought to improve their economic circumstances (often through the re-introduction of market mechanisms), more general processes of denationalisation and the limiting of the productive role of the state (within an effort to address the losses consistently made by state-run farms and industries), the abandonment of state pricing controls and subsidisation policies and, in many cases, the introduction of Western-style electoral democracy (Utting 1992: 43–44; Fitzgerald and Wuyts 1988: 1–3). It should be noted that different regimes embarked upon the reform process from different starting points. Some had nationalised extensive portions of their economies, others had not. Some had allowed markets to operate freely in certain sectors, whilst others had attempted to set prices right across the economy. All, however, sought to limit the role that the state had previously played.

There has been a fierce debate within the literature as to how far these changes represented (1) a capitulation to market forces and neo-liberal ideology, (2) a temporary search for ways of accommodating changing international circumstances or (3) a redefinition and refining of socialist forms of economic organisation which would respond favourably to the problems of the state-led strategy. To some, then, the reform process of that time represented a real attempt to find ways of tackling the limitations of the state-socialist model without simply capitulating to the dominant neo-liberal ideology. Fitzgerald and Wuyts (1988) are, perhaps, the most eloquent supporters of this view. To them, the age-old debate between capitalist markets and socialist planning is misguided in the Third World context as, in the clear majority of socialist experiments, planning has proved illusory (markets have generally continued to operate, drawing scarce resources out of the formal planned economy and further reducing the capacity of the state to plan effectively). Thus, they argue that the central dilemma is not whether the economy should be centrally planned or left to market forces, but rather it is the question of how markets are to be best managed institutionally, so as to assure the most effective mobilisation of economic surplus in support of clear progressive social agendas and strategic investment priorities – whilst also allowing for the maximum scope for local initiative in the setting of those agendas and their achievement (Fitzgerald and Wuyts 1988: 6–9; see also Hodgin 1998).

Despite the possibility that the reform process embarked upon at that time could have constituted a rejuvenation of socialist economic approaches, subsequent experience has not dealt favourably with these experiments. In practice, in many cases, there was little to distinguish this reformed socialism from other attempts to ameliorate the impacts of neo-liberal economic reform through selective interventions and, in some cases (such as the Chinese experience we have already touched upon), it appears even less convincing. Furthermore, as White (1983: 20) argues, attempting to find a limited form of market regulation can be incredibly difficult since 'the relationship between them (markets and planning) is also contradictory and, without a well-conceived programme of policy reform, the results can be the worst rather than the best of both worlds – an unproductive co-existence of inaccurate planning with "anarchic" markets'. In some cases, it is incredibly difficult to assess the impact of the reform process, as policy was almost conceived 'on the hoof' in response to a range of damaging external circumstances. The Nicaraguan transition under the Sandinistas, for example, whilst endorsing a variety of ownership forms, attempted at first to administer the economy through a significantly enhanced level of state intervention. A change of direction with a limited reintroduction of market forces and much greater attention to the needs of specific sectors in 1985 was, given the escalating conflict against the Contras and the spiralling economic crisis, never really given a chance and the continuing economic crisis eventually resulted in stringent austerity measures in 1988 and 1989.

Cuba has perhaps least wholeheartedly gone down the road of market reform but, even here, there have been considerable changes over recent years as the Cuban economy has been rocked by the collapse of the support that it had received from the Soviet bloc. The need for foreign exchange has led to the search for new sources of export revenue through, for example, the development of the biotechnology, pharmaceutical and medical equipment sectors and most importantly tourism – the active promotion of which has had a profound impact on Cuban society (Pastor and Zimbalist 1995). There have also been a range of alternative approaches towards the agricultural sector which have centred around the need for food self-sufficiency and the fascinating development of organic agricultural techniques (again related to foreign exchange constraints). Nevertheless, whatever the innovations arising out of its current economic dilemmas, it is the lack of space for dissent (and hence to a certain degree innovation) within the political sphere that serves to limit its apparent wider applicability. As Burbach *et al.* (1997: 20) argue, 'the Cuban variant of socialism may survive into the foreseeable future, but until the political system opens up, the revolution will remain in a largely defensive position, unable to provide inspiration for a renewal of socialism in the Americas'. This reflects the need for a renewal of the basic model underlying the economic programme; as NACLA (1995: 6) put it, 'the resultant overlay of the old state-socialist system and the new market and cooperative forms of production has spawned a plethora of contradictions. A new socialist paradigm – which would make sense of it all and act as a compass for the future – is still sorely lacking.'

Whatever the rationale behind the reforms, then, the basic problem with the attempts at reorientating the statist bias of socialist economic strategy was that, by and large, they were not associated with a clear ideological programme and were thus interpreted as precipitous of a gradual abandonment of socialism and the embracing of one of the variants of capitalism. For instance, whilst the Vietnamese opening to market forces is still heavily regulated by the state and, to a certain degree, insulated against the world economy, there does not appear to be a long-term strategy for dealing with the impacts of market liberalisation. Indeed, as Bezanson (1998: 43) suggests, 'the debate, even at the highest levels of the Communist Party, seems to centre not on whether to liberalise further or open up the economy to the competitive forces of globalisation, but rather on how fast to do so'.

So where does all of this leave the search for a socialist alternative to neo-liberal adjustment? The first point to make is that it is generally accepted that the old state-centred approaches to socialism do not provide a workable alternative. Those socialist regimes that remain have experimented with a variety of reforms to that model but they all basically suffer from two fundamental problems. First, the fact that socialism has been so associated with state-centric ways of envisioning itself that the attempt to reform has, more often than not, ended up in some kind of market-based reform and rapprochement with the globalising international economy. Second, the traditional one-party political system with which these regimes have been associated are basically part and parcel of the old state-centred model of socialism and, as such, have proved incapable of escaping far beyond its confines. Elsewhere, there remains something of a confusion over what a more radical, less state-centred, socialist alternative might look like. In some ways, this is a reflection of a change in the left's understanding of itself. Today, most left-of-centre political movements conceive of the constructing of a new, more egalitarian, society as something that will only take place over a long and extended period of time. This view suggests the abandonment of a long-held (and perhaps inherently debilitating) trait of twentieth-century socialist thought, 'the belief that socialism means redemption, salvation, the cure of all ills that have plagued mankind since the beginning of time, a world made anew with new men and women' (Miliband 1996: 18).

Does this mean that Castañeda was right after all? Is the best that we can hope for a long-term struggle to give 'capitalism a human face while maintaining a social order dominated by it' (Miliband 1996: 19) – a gradual reform of neo-liberalism from within? There are certainly many leftist political movements that appear to have answered 'yes' to that question – others, clinging to traditional dogma, reject its relevance. Still others are, however, more cautiously optimistic. Burbach *et al.* (1997) argue, for example, that the dearth of realistic alternatives to neo-liberalism needs to be explicitly recognised, leading to the acceptance of Castañeda's piecemeal reformist approach to national politics and all of the compromises that will involve. Nevertheless, they argue that such a strategy must be guided by the long-term goals of enabling the emergence of a new economic model through protecting and sustaining new economic subjects amongst the majorities marginalised by neo-liberal globalism and fomenting the transformation of civil society and local political cultures in directions that might challenge the discursive dominance of neo-liberal ideas. As they put it, 'in the short term reformism is the political scenario that the left has to work within as it struggles to articulate and consolidate a new historical project ... these parties will not dramatically change their countries, but they may shift the balance of political debate and provide space for new ideas and approaches' (Burbach *et al.* 1997: 166–167). These themes are explored in detail in the final chapter.

9 Social movements, the state and civil society

Ed Brown

From both a leftist and a rightist perspective, the state [is now being viewed] as an instrument of exploitation, pre-empting popular or individual initiative. As the pendulum swings in the opposite direction, analysts now maintain that developmental wisdom is lodged not in government bureaucracies but in local communities and institutions. Indigenous knowledge and popular participation are examples of concepts that have come to occupy increasing prominence in the debate.

(Hyden 1997: 4)

This final chapter shifts attention away from both wider international contexts and national political struggles for state power, to consider the possibilities that more effective alternatives to neo-liberal adjustment may grow from the local strategies of resistance of a range of social and political groups in particular localities. Such perspectives are often, as suggested by Hyden above, linked to positions which share the neo-liberal distrust of the state and a conviction that it is only in the initiatives of local communities and the wider organisations of civil society that real advances in democratisation, empowerment and political participation will be achieved. This championing of the local over the national, civil society (however defined) over the state and the informal over the formal (in terms of political participation) has won many adherents amongst the left who, disillusioned with the top-down political styles of socialist regimes in the past, have interpreted the impressive range of so-called new social movements as a form of wider popular resistance to neo-liberalism and possible 'building blocks' for the construction of wider political alternatives.

Nevertheless, the increasing attention paid to small-scale local economic initiatives, the organisations of civil society and, perhaps of most significance, the massive and rapidly growing number of non-governmental organisations (NGOs) of both North and South, has not been limited to those on the left seeking a political alternative to neo-liberalism. To some, these initiatives are important in and of themselves, rather than through their potential linkages to wider social and economic transformation. This reflects a growth of interest in a range of local economic activities, largely reflecting a 'self-help' philosophy, where members of the local community have formed their own credit unions or pooled their resources in other ways. These types of initiatives have been increasingly advocated and facilitated by international aid agencies and NGOs who, despairing of the impacts of adjustment programmes, have seen such local provision as offering some sort of ameliorative effects. Over recent years, however, such activities have also begun to find favour within the World Bank itself, as it has seen their potential for acting as a pressure valve to ease the passage of the removal of public services. Furthermore, as we saw in chapters 5 and 7, the gradual replacement of the state by new 'modern' private organisations, the ethos of self-help and the facilitation of micro-enterprise, fit well with the IFIs' new focus upon market-friendly intervention. As such, local self-help organisations are increasingly seen as effective and efficient ways of responding to the vacuum created by the gradual stripping back of the capacity of the state to provide social services. The ironies provided by the diverse range of political actors advocating local solutions are illustrated well by Judith Hellman (1997: 14–15) who suggests that 'social movements are simultaneously acclaimed by enthusiasts at opposite ends of the ideological spectrum as an expression of popular resistance that may rescue the world (or at least movement participants) from the predations of neo-liberal policies and as a tool through which neo-liberal progress can be made to work more effectively'.

It is this shared fascination with, and promotion of, local economic solutions and socio-political actors and, moreover, the sharply divergent motivations underlying their advocation, which forms the focus of this final chapter. Before entering into some of the debates surrounding these issues, however, some definitional clarification is needed. Not all commentators mean the same thing when they talk of civil society, social movements or even participation – a factor which reflects the range of ideological perspectives hinted at above – and such terms are bandied about in the literature somewhat loosely and interchangeably. Thus, in some analyses, the designation 'social movement' is used to refer to any associative grouping. In others (often where prefaced with the word 'new') it explicitly excludes traditional political actors, such as political parties and trade unions, and includes pretty well everything else from environmentalists to basket weavers. Whilst in still others, it has a much more expressly political meaning. Similarly, civil society can be a difficult term to pin down. As pointed out by Allen (1997: 330) its oldest meaning is as a way of conceptualising society when it becomes 'self-consciously politically active'. Recent years have, however, seen a vast range of different approaches to civil society which have differed according to whether the term is primarily reflecting economic or political activity (business associations or political movements) and whether civil society and the state are autonomous spheres or organically linked (Allen 1997; Hyden 1997; McIlwaine 1998). Cox (1999: 7–8) helpfully suggests that the usage of the term essentially boils down to two juxtaposed meanings. The first reflects an essentially 'top-down process in which the dominant economic forces of capitalism form an intellectual and cultural hegemony which secures acquiescence in the capitalist order among the bulk of the population'. The second, on the other hand, 'envisages a bottom-up process led by those strata of the population which are disadvantaged and deprived under the capitalist order who build a counter-hegemony that aspires to acquire sufficient acceptance among the population so as to displace the erstwhile hegemonic order'. This is a distinction which is returned to below.

Here we explore several debates relating to these issues in the light of the differences in definition and focus that we have briefly touched on here. We begin by considering the growth of local grassroots economic projects, before going on to consider the phenomenal growth in NGOs and their significance more directly. This then moves on to a discussion of the politics of social movements and their relationship with the traditional left, before finally linking back into the wider national and international contexts.

Grassroots economic strategies

A good starting point for exploring these issues is provided by a brief consideration of the, largely NGO facilitated, growth and advocation of local grassroots development strategies – which have essentially taken the form of discrete projects in such fields as credit provision, agricultural extension services, self-build housing programmes, etc. Whilst these types of projects have long been part of the development industry's portfolio, recent years have seen them take an increasingly prominent part in development financing and they are often now seen as a form of development strategy in themselves. Most of these initiatives are geared towards meeting the basic economic needs of individuals and/or communities and are generally facilitated through NGOs, either reflecting co-operation with independently formed local organisations attempting to find 'spaces' for their economic survival or directly by the NGOs themselves. Whilst these types of projects are extremely diverse in their nature and forms of organisation, a whole mini-paradigm of grassroots, neo-populist or alternative 'development' has grown up around the championing and facilitation of these types of local development strategies. Central to this form of thinking are the stress upon self-reliance, the re-establishment of the community (or, indeed, the individual in the more neo-liberal-inclined organisations) as the basic subject of development and strong emphases upon ideas of sustainability and popular participation.

Many of these concepts have found their way into the discourses of the IFIs over recent years, as they have stepped up their funding of NGOs working in many different sectors – reflecting both a recognition of the role that such projects can play in ameliorating the worst social impacts of adjustment and their desire to deepen transformations in the relationship between civil society (often interpreted as reflecting private economic actors) and the state. The crudest neo-liberal positions see the poorest sectors as a 'breeding ground for entrepreneurial innovation' – the key, in their minds, to cultural 'modernisation' and development (such ideas are frequently based on the influential work of Hernando de Soto 1989). The influence of such thinking on the IFIs' financing of grassroots projects has led to an emphasis within the increasingly professionalised

NGO sector upon the individualistic 'self-help' philosophy of micro-enterprise, rather than the community-development rationale which had previously dominated. By and large, the advocation of targeted initiatives (be they from the state or the NGO sector) aimed at facilitating that entrepreneurial spirit is clearly illustrative of the recent recognition by the IFIs of the need for action to overcome some of the structural impediments that can impede the free operation of markets (in this case often the rural financial markets) through limited market-friendly forms of intervention.

Other projects continue to be more community-based but most share a kind of blindness towards the wider political and economic context within which local self-help strategies must be situated (projects designed, for example, to facilitate the co-operative production of goods for local markets must face the challenges of operating in liberalised markets where import competition is incredibly intense – Carmody 1998: 34). Nevertheless many of the advocates of grassroots economic contexts see their lack of articulation with wider economic and political programmes as an advantage rather than a limitation. They are not adverse to seeing such initiatives eventually lead to an alternative set of economic proposals but have generally abandoned the active search for such strategies as largely self-defeating and unlikely to significantly affect the lives of people in individual communities. In a sense, the advocation of these projects as a kind of grassroots alternative to neo-liberal adjustment is the other side of the 'there is no alternative' neo-liberal scenario. In other words, it is a strategy that appears to accept that there is no workable alternative for national economic management, leaving the possibilities for experimentation and improving people's lives as a matter of local initiative and private charity. Of course, this somewhat overstates the case and is, furthermore, far too dismissive of important innovations which have, on occasion, offered hope, dignity and resources to people in often extremely marginalised and inhuman conditions. There are opportunities for experimentation at the local level which may simply not be possible at wider levels given current international realities, and which may prove the building blocks of new, locally generated, economic strategies. As Duncan Green argues, 'starting at the grassroots also tends to move the discussion away from the search for one "big idea" with which to oppose neo-liberalism'. Nevertheless, he continues, 'the "small is beautiful" approach does entail political costs for those seeking alternatives. ... As anyone who has lived through the Thatcher–Reagan years will know, the endless repetition of a simple, intelligible and apparently coherent "big idea" is a far more powerful political weapon than saying "There are no simple solutions, we'll have to look at everything case by case"' (Green 1995: 194–195).

Some of these issues can be effectively illustrated through a brief discussion of micro-credit schemes. These schemes have recently been the subject of substantial international attention from a wide range of organisations who view them as an increasingly important weapon in the fight against rural (and indeed urban) poverty. The growing advocation of micro-credit culminated at the three-day Microcredit Summit in Washington during February 1997 (see the

Summit web site at http://www.microcreditsummit.org/ for further details) which brought together a broad range of NGOs, international financial institutions (including the World Bank) and the representatives of international financial concerns, such as Chase Manhattan and American Express. Whilst this reflected agreement amongst such organisations as to the importance of micro-credit schemes, the motivations that underlie their support are more varied.

NGOs began advocating micro-credit schemes in the early 1980s as a way of enabling the poorest sectors to attain access to financial resources that might finance activities through which they could work their way out of poverty. This reflected the failure of market liberalisation to improve the poor's access to finance. In fact, it had frequently made matters worse through the escalation of interest rates and the even greater unwillingness (in comparison to nationalised banks) of private financial organisations to deal with the poor. The micro-credit schemes attempted to target donors' assistance much more directly to the most deprived sectors through savings and credit groups (specific groups were targeted rather than spending categories) and self-selection was encouraged through limiting the size of loans and demanding attendance at regular group meetings (Marr 1999: 2–3). Gradually, a number of NGOs began to expand their programmes and eventually governments and the IFIs began to see their potential. Official financial policies towards the poor, thus, increasingly began to shift from the subsidisation of interest rates to providing assistance to the NGOs running micro-credit schemes (through direct finance, training programmes, the setting up of governing bodies and the provision of financial incentives to staff – Marr 1999: 3). The IFIs have not only supported these schemes because they reflect their emphases upon private sector development (whilst still meeting some of the criticisms of the social effects of their wider structural lending), but also because, along with the private banks, they have seen them as potentially extremely profitable. The World Bank signalled its enhanced interest in the whole area of micro-lending through the recent creation of its Consultative Group to Assist the Poorest (CGAP – see material on the World Bank web site) which, whilst relatively small in relation to total Bank lending, could prove enormously influential in the handling of the likely growth of this type of activity (see for examples the guidelines already issued in World Bank 1995). The rapid growth of these types of initiatives and their evolving relationship with the IFIs and national financial markets have tended to produce an increasing professionalisation of the micro-credit phenomena. This tendency has already begun to attract its critics. As Rahman (1999: 79) puts it, the most important criterion for success of micro-credit programmes is being increasingly 'determined by their ability to achieve financial stability ... [but] ... the service-providing institutions must also consider whether the attainment of such sustainability involves too large a cost in terms of borrowers' socio-economic impoverishment'.

Micro-credit and other local economic schemes do not, of course, only have their supporters amongst the more politically conservative sectors. Indeed, as we have already pointed out, the majority of such initiatives originated amongst those disillusioned by both the impacts of the neo-liberal adjustment process on the vast majority of the poor and the seeming inability of any alternative ideas to significantly reach those groups either. Many NGOs view the schemes as an opportunity to pursue programmes that reflect their concerns for community participation, economic security and empowerment of the poorest. Micro-credit schemes have certainly improved the possibilities for individuals and communities to grasp new economic opportunities. Their record on the less tangible measures of empowerment and social cohesion are less easy to pin down; although it would appear that as schemes get bigger and more professionalised, their ability to work towards those goals decreases. In effect, the major conclusions of a growing number of sceptical observers are that, whilst there have been some important successes, the major problem with micro-credit schemes is that they have failed to adequately provide for unexpected fluctuations in the incomes of the poorest borrowers, that they have not been sufficiently linked into programmes designed to guide usage of the resources provided and, finally, that they may be diverting resources away from other activities with greater potential impacts upon poverty (Marr 1999: 4). Furthermore, it is also worth remembering, in the words of Judith Hellman (1997: 18), that, whilst 'micro-enterprises may develop into democratically managed cooperative projects', they may also become 'a further source of inequality and stratification – not to mention tension and violence – in poor communities'. Even the often-praised Grameen Bank in Bangladesh has come under criticism recently, through studies which have revealed something of a gap between the ideology and vision of the bank (i.e. the way in which it presents itself to national and international communities) and its practices in the field (Rahman 1999: 79; for other recent discussions of micro-credit see Singh *et al.* 1997 and Fernando 1997).

The NGO-isation of the world

The issues raised from our brief discussion of micro-credit lead us to a somewhat wider topic – the significance of the explosion in the size, influence and number of NGOs over recent years. Their growth in numbers, the sheer weight of the financial resources at their disposal and their increasing involvement with the IFIs have combined to produce a situation where, today, individual NGOs may be more important political and economic actors in the regions where they operate than local government. NGOs now have a significant presence in social welfare provision and employment creation and have also, in recent years, received an increasing proportion of IFI project expenditure as well as considerable access to the decision-making process within the institutions (Hulme and Edwards 1997: 3). USAID, for example, announced in 1994 that they were intending to administer half of their financial assistance to the South through private sector organisations. Whilst, it is difficult to estimate the total number of NGOs now operating across the globe, there is little doubt that they are of considerable, and growing, importance in many spheres of activity (Stewart 1997: 11).

Despite their obvious significance, there are, nevertheless, a wide variety of different organisations which we might encompass under the general heading of NGOs. Again, there is some definitional confusion in the literature. For some, the term NGO refers to any non-profit organisation – whilst others make a distinction between NGOs ('intermediary organisations engaged in funding or offering other forms of support to communities and other organisations that seek to promote development') and grassroots (or popular) organisations which are, generally smaller-scale, membership organisations (Hulme and Edwards 1997: 21). Following this distinction, we can further sub-divide NGOs themselves into Northern and Southern organisations, with the former sometimes administering projects directly and sometimes working with their Southern counterparts or grassroots organisations in a variety of different arrangements. Northern NGOs themselves differ considerably in terms of their size, goals, relationships with local governments and international funding agencies and the types of activities which they carry out. Southern NGOs are also diverse with some working at a relatively wide scale across whole countries (and sometimes beyond) and others which are much smaller and tend to work within particular localities. It can actually be very difficult to adequately distinguish between local NGOs and popular grassroots organisations which also sometimes receive access to external funding (McDonald 1997: 6–8). This great variety of different types of NGO can make an analysis of their impact very difficult. For our purposes, within this section on the wider significance of NGOs, most attention is placed upon the role of the larger international NGOs, whilst the following section deals more directly with the smaller, more autonomous, organisations.

The inexorable rise in the number and importance of NGOs has been viewed from a range of different perspectives. The IFIs have gradually shifted from a position where NGOs were mistrusted and seen as a challenge to the pursuit of appropriate economic reform to one which now views NGOs (or at least those which most closely relate to the newer more socially responsible version of neo-liberalism) as important partners in the transformation of developing countries. Many of the opponents of neo-liberalism have also been attracted to NGOs. Jenny Pearce (1997: 257), for example, refers to a 'ripple of excitement' with which progressives in Latin America greeted 'the discovery that the negative impact of the retreat of the state ... could be moderated by the dynamism of the myriad of organisations outside the state committed to grassroots development'. Much was made of the apparent abilities of NGOs to work at local levels with local organisations, highlighting the democratic credentials of such approaches over the centralised and coercive approaches said to characterise the much maligned Third World state. NGOs were seen as cheaper, more flexible and more participative and also more likely to achieve transformations that would strengthen civil society and encourage democratic participation, the rule of law and citizenship. Nevertheless, as the significance of NGOs has grown, and more and more resources have been placed in their hands, they may have begun to lose the characteristics claimed for them. As Stewart (1997: 13–14) argues:

The positive characteristics ascribed to NGOs are the characteristics of small organizations – flexibility, lack of bureaucracy and closeness to the target group to name three. Small organizations may intervene extremely effectively in their own constituencies, but it is clear that they cannot meet the challenge of enormous underdevelopment. Hence, scaling up – a theory which sets to show that small organizations by growing themselves in a number of different ways ... can deal with large-scale developmental problems whilst retaining their small organization charms. ... Finally, there is little evidence that small NGOs when they grow large retain their small organisation characteristics. Research indicates that as their size increases, NGOs experience all the difficulties associated with large organisations. ... A further problem is that, as NGOs grow larger, their need for funding increases and this creates institutional schizophrenia, with fund-raisers focusing on their constituency (donors), whilst field staff focus on theirs, the target group.

Recent years have, therefore, seen a growing questioning of the progressive role assumed for NGOs. As Laura MacDonald reminds us:

in searching for alternative models of development ... progressives need to question the assumption that NGOs are necessarily allies in a common cause. When it comes to aid, small is not necessarily beautiful. In fact assistance from Northern NGOs to small development projects can be even more dangerous than aid from state agencies because it penetrates into the very fibres of a community, creating new forms of clientalism and co-option.

(MacDonald 1995: 35)

There has, in particular, been a sustained critique of the democratic credentials of NGOs. Unlike the institutions of the state which have, at least in principle, some accountability to local electorates, NGOs are often answerable only to those who provide them with their financial resources. It is worrying that, despite a commitment to the ideals of empowerment and participation on the part of most NGOs, local communities may have little impact upon the decision-making processes of what are, in effect, the most important economic and political actors within their region. Furthermore, as NGOs become a more and more central component of the dominant development agenda (and hence increasingly market-driven) there is a danger that the smaller more locally accountable and directly participative organisations may be forced out (Potter *et al.* 1999: 181). A point which leads Najam (1998: 305–306) to ask whether the increasing popularity of NGOs in 'official' circles reflects a genuine recognition of the importance of the NGO's original focus upon local accountability and empowerment or, rather, the fact that, gradually, NGOs have been 'socialised' into the development industry establishment. Thus, the rhetoric of participation, so clearly associated with the NGO sector, is not necessarily all that it might appear. As Kothari (1984; cited in MacDonald 1996: 204) puts it, 'the more the economics of development and the politics of development are kept out of reach of the masses, the more they (the masses) are asked to "participate" in them'.

This leads us to a further point which relates to the continued determination of development agendas by external forces. Many have proclaimed the local-level focus of NGOs as offering a real opportunity for local communities to more actively determine their own futures. However, on closer reflection it is clear that, in reality, NGOs can be (1) as constrained by the whims of their funders as governments and (2) as prescriptive in their dealings with local communities as state institutions were accused of in the past. As such, the types of activities which they fund (which may represent considerable flows of resources to a region which will carry significant local political impacts) will probably (despite the rhetoric of participation) owe far more to the dictates and trends of external funders, than to the desires and necessities of the beneficiaries themselves.

Perhaps the most important issues, however, revolve around the relationship between NGOs and the state. Many opponents of the neo-liberal agenda initially saw the burgeoning NGO sector as entering into a healthy alliance with the state, with NGOs providing an indispensable role in filling the gaps in social provision left by the retreat of the state. This suggested an alliance between the NGO sector (sometimes expressed more widely as civil society) and the state with a new, and more efficient, distribution of resources between the two. Gradually, however, the relationship between the NGO sector and the state has been seen as much more antagonistic than this would suggest, with NGOs seen as supplementing and, in many cases, replacing the state and generally opposed to, or disengaged from, state policies and strategies, rather than complementary to them (Marcussen 1996: 406). Given the record of many political elites and governments in the South this might not necessarily be seen as a problem by progressive sectors. For one thing, the state can obviously not carry out all of the tasks that it used to and NGOs may well be able to carry out those tasks more effectively and transparently – although there is certainly no automatic principle that would guarantee that. In fact some case studies of the role of NGOs cast serious doubts upon their efficiency. As Rocha (1999: 43) explains in relation to the role of NGOs in San Francisco Libre, Nicaragua:

> It is very common for NGOs to know nothing about the subject with and for whom the work is being done. The most pathetic case of this in San Francisco Libre is the promotion of nontraditional crops in a municipality that lacks even a minimum transportation flow. ... NGO project officers would do well to ask themselves when was the last time they slept in a peasant's house. The answer will explain how an NGO can wax euphoric about San Francisco's rice potential, basing its theory on the area's superficial humidity, when the reality, according to local residents, is that the water table in most of the municipality lies seven meters underground.

He goes on to catalogue a whole host of problems involved in NGO interventions within the municipality, including (1) the fact that projects are only supported if they fit donor-identified priorities, (2) that there is often little local input into the design and implementation of projects, and (3) that some localities have a range of different NGOs working within them that duplicate projects (or even actively counteract each other) whilst there may well be no NGOs at all five miles further up the road. Where, in such circumstances, are the chains of responsibility? Where is the co-ordination of developmental efforts? Nevertheless, at least the NGOs are there and, as Rocha (1999: 44) himself so succinctly puts it, 'even a bad NGO is better than an absentee government, "There's plowing to be done, and these are the only oxen we've got".'

One is left wondering quite where the responsibility for national development now lies – a concern which has led some commentators to an even more sceptical view of the growth of the NGO sector. Hearn (1998), for example, explicitly connects the courting or co-option of the NGOs to the most recent stage of the neo-liberal adjustment project. He identifies the incredible evolution of the NGO sector and the supplanting of the state as a form of social engineering, whereby the nature of developing countries is being consciously transformed through the actions of development agencies. As Frank Judson (1993: 162) puts it, the notion of economic crisis and poverty is in the process of being removed from economic decisions and the actions of the community as a whole or, in the final instance, the state – to become a matter of individual responsibility. He carries on:

> If economies are efficient in 'market terms', then states are successful; in a universe of entrepreneurial initiative and individual responsibility, a 'crisis' of poverty and marginalisation is an individual matter, at best a matter for 'civil society' to take up, i.e. charity and self-help, not an arena of public policy and state responsibility.

The left and the grassroots movements

We concluded our discussion of NGOs by suggesting that their growth, whilst certainly offering some opportunities, is not all that it might seem. Nevertheless, there are a great variety of NGOs and many (especially the smaller and more independent) have retained a strong relationship with autonomous social movements and local communities. Similarly, the growing focus upon the transformation of civil society and the facilitation of local grassroots projects does not, of course, only relate to the enhancement of entrepreneurial skills and the strengthening of the private sector *vis-à-vis* the state. Many sectors of the left continue to see the promotion of the organisations of civil society (including NGOs) as the most effective possibility for the pursuit of social transformation. Here, it is worth returning to our earlier definitions of civil society, where we distinguished between a top-down conception of the strengthening of civil society that served to legitimate the existing order and a bottom-up approach which attempted to build a counter-hegemony – an alternative understanding of society and citizenship which would arise to challenge the capitalist order. Such a distinction can also be applied to different types of NGO (often interpreted as the major constituent organisations of civil society) as follows:

> What is the fundamental difference that distinguishes an NGO with a popular orientation from an NGO linked to the Neoconservative strategy? Both implement small projects, they both link themselves with the most vulnerable social groups. ... In reality what distinguishes a Neoconservative wave of NGOs from an NGO movement committed to promoting the leadership of popular groups lies in how they view the problem of power. In the first case, the activity of the NGO is oriented at promoting changes in order to avoid modifications in the structure of power. In the second case, the NGOs try to promote changes in order to achieve transformation in the relation of social forces, in a manner which favours the majority.
> (Concertación Centroamericana de Organismos de Desarrollo 1988
> *Memoria: Reunion de Organismos No-Gubernmentales para la
> Constitución de la Concertación Centroamericana de
> Organismos de Desarrollo*, San Jose, pp. 24–25,
> cited in MacDonald 1996: 211)

In reality, however, whilst the left has embraced the role of NGOs with a popular orientation, the major focus has been upon the vast range of associational social and political movements that have emerged (or perhaps been discovered) in the South over recent years. These movements are 'action', rather than 'project', oriented and function through the participation and mobilisation of their members rather than via rigid or hierarchical institutional structures or the inflow of financial resources. As Sinclair (1995: 5–6) argues, 'popular movements may develop institutional structures – such as non-governmental organisations (NGOs), political party apparatus, think-tanks, lobby groups – but the power of the movement rests in its ability to mobilise people rather than in the power of the institutions'. The left has been drawn to such movements because of their more open resistance to neo-liberal adjustment and democratic exclusion and their anti-hierarchical and participatory forms of organisation. Nevertheless, as was highlighted in the introduction to this chapter, there are a massive variety of movements and considerable diversity in the claims made for their potential progressive impacts. One of the most frequently recurring claims made for these organisations is that, together with the more popularly inclined NGOs, they may represent the emergence of new social forces which may, given time, lead to the articulation of a new progressive political project linked to new economic rationales. Sinclair (1995: 15), for example, argues that:

> within the informal economy and within local communities there exists greater space to propose, tinker with and build economic alternatives.

Production co-operatives, community savings and loans associations, job training and literacy programmes and many other alternative ideas have greater chance for success in local communities where people know and trust each other.

There is, nonetheless, a recognition that, at present, whilst there is a growing conscientisation occurring within these sectors and some local successes in experimenting with new grassroots economic strategies, there is an urgent need to move from protest and local experimentation to concrete collaborative proposals, with well-defined political programmes that can begin to challenge the current discursive hegemony of neo-liberal ideas. For some, the development of any such wider movement is likely to be an extremely extended process and one that will be based far more in the gradual transformation of civil society and dominant value frameworks and political cultures than in the formation of political alliances geared towards the capture of state power. In either case, however, many would argue (although this is certainly not a view shared by all commentators) that for grassroots movements to have any significant impact upon the broader context within which they function a clearer ideological presence and closer co-operation between distinct types of organisation is essential (although whether that process should reflect an increasing concern to influence the state or engage with more traditional elements of the left, such as political parties and trade unions, is hotly debated).

For others, however, grassroots social movements are far too diverse to offer any real possibility for wider collaboration or a more unified national political presence (an indication of the fragmentation imposed by late-twentieth-century capitalism). Yet, at the same time, they also celebrate that diversity – seeing the new social movements as an appropriate form of resistance to the dominant forms of cultural control through their struggles to establish and protect autonomous identities within each of the fragmented spaces of that system. For example, authors such as Escobar and Alvarez (1992) would argue that the new social movements have very different objectives from the traditional class-based movements of the left and, therefore, should not, and indeed cannot, be subsumed into some new attempt at creating a socialist utopia. In their view, then, social movements are not primarily, as has been inferred above, forms of resistance to changing material and political circumstances (or demands for wider political transformation) but rather they reflect local struggles for the 'spaces' to be able to project autonomous definitions of self and community. To such theorists, the importance of these movements is to be found within their struggles to challenge and, moreover, expand the dominant understanding of what is deemed as 'political' – the questions of identity, of behaviour, of the very language of modernity, tradition and development (Shefner 1995: 605). In this sense, the potential of grassroots social movements lies not in their potential role in invigorating or redefining traditionally defined political struggles (or the economic opportunities which they may embody) but, rather, in the impacts that they can have upon the nature of civil society – the production of changes in people's view of themselves and others, of their local community and its place in the world, of their values and potential life-choices – as a necessary prelude to any wider social transformation. In this way, the nurturing of grass-roots social movements is seen by such authors as a potential discursive challenge to the dominant values being articulated within the current focus of the IFIs (and their new partners in parts of the NGO sector) upon the construction and transformation of civil society. The growth of social move-ments is, then, seen as a diverse autonomous process for which there can be no pre-conceived ideological aims – as Arico (1992: 22) puts it in relation to the redefinition of any socialist alternative, 'we ought to think of it as a process for changing people's mentality. Socialism is, thus, a counter-cultural force for changing culture, rather than one that seeks particular goals.'

Others, even if they view the movements as having a more traditional polit-ical role, still see their major potential as working towards the production of changes in the dominant value frameworks within society and enabling the emergence of a more coherent and radically different body of political thinking. Burbach *et al.* (1997: 158), for example, argue that:

> many of the leaders of these movements even question whether it is appro-priate to hold state power at present, understanding the need to accumulate more forces, to develop more coherent ideas and values that can really change society and the global economy.

They continue:

> communism failed in part because it was born prematurely. The same mistake shouldn't be made again, of launching a project of state power that will be still-born or aborted because the full-blown elements do not exist yet for building a new society.

For others, the whole focus upon grassroots movements and, in particular, the view that they represent starkly new forms of social and political activity is somewhat misplaced and based upon a misreading of social responses to the impacts of global economic change and transformations within dominant polit-ical cultures. Questions have been raised regarding the silence within much of the literature on the abilities of the state to co-opt or crush autonomous move-ments and the debilitating hostility towards traditional formal politics as the major arena for influencing the activity of government, and the sheer diversity and differing political philosophies of the movements and their relationships with the NGO sector. Thus, for some, an appropriate appraisal of the potential of new social movements in aiding the construction of an alternative to neo-liberal adjustment requires a more selective and sceptical approach which moves beyond a blanket support for their activities. In particular, authors such as Petras (1997a and b) and Veltmeyer (1997) propose a more traditional class analysis of the emergence and activities of social movements which retains an understanding of 'social movement action as purposive political action, and not just the autonomous expressions of cultural constituencies' (Shefner 1995: 605). Hence, Petras (1997), for example, points to the particular importance of a new wave of rural movements which have emerged in Latin America over the past decade. He explores their emergence in relation to the changing political economy of Latin American agriculture and the gradual drift towards electoralism and opportunism of many sectors of the traditional leadership of the left, as well as the considerable organisational and ideological innovations of the movements. Despite their differences from traditional peasant associations (their greater links to the cities and to wage labour and their greater political autonomy), he argues that they are still fundamentally class movements:

> such groups are not simply 'new social movements'. They retain and develop Marxism in new circumstances adapted to new class actors engaged in novel types of struggle with the clear perspective of changing the national, if not international, structure of political and economic power ... the fundamental turn that must be made from strictly agrarian issues to social transformation has to be built around the renewal of a socialist praxis that links cultural autonomy and small-scale production with control over the commanding heights of the economy.
>
> (Petras 1997b: 43, 47)

There are, therefore, a wide range of positions on the potential role of grass-roots organisations to aid the articulation of a radical alternative to neo-liberal adjustment – there is not the space here to explore all of these in detail. We conclude this chapter with a brief consideration of one of the most influential recent interventions within the literature – the work of Roger Burbach and his associates in *Globalization and Its Discontents* (1997) which was previously discussed. Here, as the reader may recall, the authors argued for a two-pronged strategy that combined the slow nurturing of a range of new social actors arising from the economic activities engaged in by those marginalised sectors left out of the exclusionary neo-liberal economic model with engagement in formal political struggles to ensure the economic and political conditions for the maturation of the new sectors. They refer to these new sectors as post-modern economies – the diverse economic activities engaged in by the underemployed and the discarded sectors of society. They argue that, over time, the numbers forced into such activities will increase as neo-liberal globalisation produces ever-more stratification and marginalisation. As they explain it:

> Late capitalism does not have the capacity to absorb these petty producers in any significant way. In fact their numbers will only swell as globalization, driven by ever more advanced forms of technology, requires fewer and fewer workers to turn out goods and services. The dominant economic system can certainly function without the excluded, indeed it discards them

as redundant and useless. But there is another side to this rejection in that it turns the excluded into discontents who have no use for the dominant system. It gives them nothing and makes them potential insurgents.

(Burbach *et al.* 1997: 155)

They argue that these economic activities will begin to coalesce with other similar economic endeavours such as co-operatives, worker-run enterprises and municipal enterprises. Together, these constitute a potentially vast 'class' of associative producers. Thus, for Burbach *et al.*, the key to the wider significance of grassroots economic endeavours is their associative nature, their commitment to co-operation (and often community development) as opposed to the celebration of individual initiative and financial efficiency of much of the more mainstream interest in local economies. Immediately, the question arises as to why and how these associational forms are likely to develop. Burbach *et al.* argue that this will occur because there will simply not be the opportunities for these economic strata to assimilate into the wider global economy (as is the intention of most mainstream programmes geared towards these sectors). They continue:

> Furthermore, the present system of corporate capital is restraining human development and economic creativity at the bottom. In the long term small capital and the emergent system of associate producers can only eliminate the 'fetters' imposed on them by organizing, or 'associating' among themselves and engaging in a struggle against big capital.
>
> (Burbach *et al.* 1997: 156)

Their basic argument is that these new economic sectors constitute a new mode of production that will somehow rise to become the dominant form of economic activity in the future. They stress that this will be a long-term process, as the current system continues to go through its periodic crises and the new economic actors develop their own identity and political programmes. The difference, then, between Burbach *et al.*'s prioritising of these sectors and the more mainstream focus on the local and some of the left's celebration of new social movements, is that here it is attached to a political agenda committed to wider processes of social transformation. At present, that political programme remains in its birth throes – as Roger Burbach (1997: 20) argues elsewhere, the new mode is 'a socialism of place, a socialism with a local agenda, a socialism with a hundred faces and experiences, a socialism without a name or a grand narrative at present'.

The eventual emergence of this new mode of production is, of course, not an historical certainty but depends upon the political forces that will give it shape and, as we have seen, that wider political project is still sorely lacking. Burbach *et al.* (1997: 157) recognise that 'to mobilize and consolidate the diverse group of nascent producers a new ideology needs to counterpose grassroots economic development to the domination of big capital'. Who will provide the leadership for this political project? They suggest that this is already beginning to emerge from the broad range of social movements struggling to challenge the concentration of capital, certain types of NGOs (especially in the South) and the increasing numbers of professionals discarded by the system. Given the lack of guiding principles and the nascent state of the new alternative, they argue for a political programme geared towards gradual reformism that would work to guarantee the economic resources (appropriate access to credit, controls on the concentration of capital, and so on) and the political spaces (respect for constitutionality, enhanced political participation, etc.) for the long-term nurturing and maturation of such experiments. This political analysis is, perhaps, the weakest part of their position. There is little evidence that the political will to protect these spaces necessarily exists and, electorally, a long-term struggle for growth, if not survival, is certainly not going to be as attractive as a political strategy that offers more immediate gains (even if it is probably far more accurate). The political implications of their analysis certainly need more attention than can be given here. Nonetheless, it does at least provide a starting point for a healthier discussion of the relationship between the formal political arena and the struggles of grassroots movements.

Social movements and the wider political arena

It would seem appropriate at this point to review the discussion thus far. The last two chapters have been concerned with attempts to articulate more radical responses to the dominant neo-liberal adjustment model. They considered the international context within which any such efforts must be located and discussed some of the suggestions that have been made for transforming that context as a necessary first step (or accompaniment) to alternative national economic programmes. We have also considered the legacy of socialist development strategies and the struggle to articulate a new, less state-centric, alternative that retains something of the socialist commitment to equity and social transformation. Finally, we have considered a range of more diverse local-scale initiatives – exploring the significance of the dramatic growth in importance of the NGO sector, as well as perspectives which view the broad range of so-called new social movements as possible building blocks for a range of different development (or post-development) futures. What much of this material often misses, however, are the wider international and national contexts (considered in chapter 8) within which any new alternative must be formulated. The prospects for the financing of local development strategies, for the autonomous development of popular organisations and the transformation of civil society and the experimentation necessary for the formulation of alternative forms of national economic management, remain heavily influenced by (if not dependent upon) that wider context. The types of funding available for local development projects depend upon the relationship between the NGO sector and those which provide its finance and the IFIs obviously continue to have a disproportionate influence upon the wider economic context. Similarly, despite all of its inefficiencies, compromises and corruption, the state, as neo-liberal reformers will testify, cannot be wished away and will not wither away. The political dangers of ignoring this wider picture are as important to recognise as the lessons learnt over recent years regarding the need for local participation, community empowerment and the strengthening of civil society. Even if a wider 'development model' is not intended, the ability of communities/organisations to develop independently, to assert their identity and to improve the living conditions of their members, depends upon the context within which they find themselves.

Conclusion

So where have our discussions led us? What are the prospects for the emergence of a significant challenge to the neo-liberal consensus which appears to underlie the vast majority of interventions within the debates of the mainstream 'development' literature? In an excellent article that explores precisely the challenges and limitations which we have discussed here, Robinson (1998/9) proposes four conditions for the definition of a new popular alternative to the neo-liberal model. These conditions reflect the perspective which we have pursued here – that the importance of autonomous local organisations that represent a wide variety of interests and identities cannot be understated but that they must be necessarily placed within their wider national and international contexts. We conclude by exploring his four conditions as a way of linking the three discrete themes which have constituted the preceding chapters:

(a) *The existence of a 'political force' and a 'broader vision of social transformation' which can serve to link social movements together*

Robinson accepts that some would question this condition on two counts. First, in terms of its desirability. Some would interpret any attempt to exert some sort of national presence as intrinsically weakening to social movements, through the bureaucratisation and professionalisation that it would involve, as well as the necessities for compromise and distractions involved in the intricacies of national politics. Second, in terms of its plausibility. To some the very existence of the wide variety of social movements is indicative of the ways in which changes in the global economy have fragmented national social structures such that co-ordinated action around clear aims and objectives is impossible. His response to this is that 'the challenge for popular social movements is how to fuse the political with social struggles through the development of political instruments that can extend to political society (the state) the counter hegemonic space currently being opened up in civil society through mass mobilization' (Robinson 1989/9: 126). One possibility for this type of approach is that of loose associations of organisations geared around the pursuit of specific issues or in reaction to evolving circumstances. A recent example of the potential for such co-operation comes from the formation of the Non-

Governmental Emergency and Reconstruction Coalition in Nicaragua following the horrific social costs imposed by Hurricane Mitch and the government's seeming inability to respond effectively. This coalition brought together over 320 organisations (including international and national NGOs, trade unions, regional movements, small and medium farmers' organisations and other grassroots movements). The coalition incredibly produced, within two weeks, an impressive and comprehensive contribution to the negotiations between the Nicaraguan government and international creditors and donors that called for a radical turnaround in government policies that would take seriously the needs of the poorest sectors especially in the most affected rural zones.

(b) *The formulation of a viable socio-economic alternative to neo-liberalism*

As stressed in the introduction to this conclusion, it is imperative that those committed to pursuing more egalitarian alternatives do not abandon the search for alternative forms of national economic management or the arenas of formal national politics. Economically, despite all of the local contradictions and difficult international contexts that have been described, the search for new forms of economic management is indispensable, if the economic sphere is not going to be reduced to mere tinkering with the neo-liberal model as the only acceptable form of economic decision-making. As Burbach *et al.* (1997: 168) argue, it is ultimately the state that controls and directs the capital accumulation process. It, rather than any other institution or social sector, can facilitate the development of new economic actors through its control and facilitation of credit, research and its regulation of the economy. There is, therefore, still a desperate need for a more meaningful attempt at reconstituting state intervention that is more than simply actions designed to make markets work more effectively (e.g. the potential role of the state in guiding the economy towards particular priorities). Similarly, politically, since the state remains the arena for legislation and, despite the retreats of recent years, the major influence on the domestic economic environment, the struggle for the control of government remains important, even if for nothing more than the protection of democratic spaces and curtailing the concentration of capital.

(c) *The need to transnationalise the struggles of popular movements*

Robinson's third condition reflects the view that whatever successes national political strategies might have, the broader context within which developing countries find themselves will continue to have a major influence upon the options that are open. He argues that this means that:

> the real prospects for counter hegemonic social change in the age of globalisation might be a long march in the Gramscian sense through an expanding

transnational civil society, a globalisation-from-below movement that seeks to challenge the power of the global elite by accumulating counter-hegemonic forces beyond national and regional borders.

(Robinson 1998/9: 128)

Central to this is the premise that the major impact of neo-liberal globalisation has been to transcend the reach of nationally-based democratic institutions, leading to the necessity of nurturing a new level of political institutions which are more suited to the global age. These new institutional forms will not, however, simply emerge over time as the world becomes evermore closely integrated, rather, in much the same way as occurred at the national level, the recasting of the territorial boundaries of systems of accountability and the advocation of new forms of representation and, perhaps, international bills of social, economic and civil rights, will have to be struggled for (Held 1992: 32–34). As explained by Brecher and Costello (1994: 105):

Local and national government, political parties, trade unions, grassroots organizations, feminist, environmental and other organizations have all been outflanked by global corporations. ... International economic institutions ... provide few mechanisms by which they can be held accountable. There is no global government to legislate on behalf of the world's people. One starting point for a solution lies in expanding transnational citizen action.

We have considered some of the debates surrounding these issues earlier in chapter 8 although there our attention was largely focused upon attempts to reform international institutions or far-reaching proposals for changes in the way in which the entire global economy operates. Such discussions answer questions regarding 'what' changes are proposed but do little to address 'how' such issues might be raised on the international stage and still less about their chances for implementation. Such issues in some senses lie beyond the concerns of this chapter but it is worth briefly considering a few ways in which various organisations have attempted to pursue some of these ideas; although we should stress following Waterman (1996: 168) that much of the literature on these issues is problematic in that it presents a wide variety of campaigning activities and yet often fails to discriminate between them or recognise the contradictions involved. Such activities fall into a number of different categories but most involve linking particular interest groups across borders (trade unions, environmentalists, etc.) or the linking of particular identities with wider commonalties (for example, getting environmental groups, those concerned with indigenous identity and those opposed to the World Bank's economic policies, to co-operate on issues such as the World Bank's policies towards dam construction). One of the strongest areas of cross-border activity has been amongst the labour movement in recent years with the strengthening of international trade unions, international strike support (such as that orchestrated in support of the Liverpool dock workers in the UK) and sectoral linkages through worker-to-worker exchanges and the growing attempts at cross-border labour organisation (see Alexander and Gilmore 1994 and Brecher and Costello 1994: 153–60).

(d) *The need for the re-emergence of a more fully independent group of 'organic intellectuals' much more closely linked to popular majorities and their struggles*

Robinson is referring here to the increasing tendency for intellectuals to work within structural constraints and to disassociate themselves from the struggles of the marginalised and oppressed. This reflects the ideological hegemony of neo-liberal thought (and the crisis in radical political thought) although it also responds to the increasing market pressures being placed upon the academy and, no doubt, the individual opportunism of some. As Robinson (1998/99: 128) argues:

> for these intellectuals, the structural constraints set by the system have become accepted and the only alternatives put forward as legitimate and 'realistic' are those that accept these constraints. ... It is only through struggle against established historical structures that the extent to which they can become transformed is revealed. Intellectual labour as a form of social action may constrict just as it may extend the limits of change.

This brings us back to the urgent need to critique the self-validating assumptions of modern capitalism and to articulate new visions of how society might be transformed which were highlighted in the Preface to this book. We hope that our journey through the emergence and consolidation of the neo-liberal adjustment agenda will serve in some small way to destabilise some of its totalising assumptions and open up the possibility for genuine alternatives.

Bibliography

Adams, W. (1993) 'Sustainable Development and the Greening of Development Theory', in F. Schuurman (ed.) *Beyond the Impasse: New Directions in Development Theory*, London: Zed Books.

Addison, T. (1993) 'A Review of the World Bank's Efforts to Assist African Governments in Reducing Poverty', *ESP Discussion Paper*, no. 10, Washington, DC: World Bank.

Adedji, A. (1995) 'An African Perspective on Bretton Woods', in M. ul Haq, R. Jolly, P. Streeton and K. Haq (eds) *The UN and the Bretton Woods Institutions: New Challenges for the Twenty-first Century*, Basingstoke: Macmillan, 60–82.

Agnew, J. and S. Corbridge (1995) *Mastering Space: Hegemony, Territory and International Political Economy*, London: Routledge.

Akinterinwa, B. (1994) 'The Experience of Niger: Adjusting What Structure?', in A.O. Olukoshi *et al.* (eds) *Structural Adjustment in West Africa*, Lagos: Pumark.

Alavi, H. (1972) 'The State in Postcolonial Societies: Pakistan and Bangladesh', *New Left Review* 74.

Alexander, R. and P. Gilmore (1994) 'The Emergence of Cross-border Labor Solidarity', *NACLA Report on the Americas* 28, 1: 42–48.

Allen, C. (1997) 'Who Needs Civil Society?', *Review of African Political Economy* 73: 329–337.

Amoore, L., R. Dodgson, B. Gills, P. Langley, D. Marshall and I. Watson (1997) 'Overturning "Globalisation": Resisting the Teleological, Reclaiming the "Political"', *New Political Economy* 2, 1: 179–195.

Aribisala, F. (1994) 'The Political Economy of Structural Adjustment in Côte d'Ivoire', in A.O. Olukoshi *et al.* (eds) *Structural Adjustment in West Africa*, Lagos: Pumark.

Arico, J. (1992) 'Rethink Everything (Maybe It's Always Been This Way)', *NACLA Report on the Americas* 25, 5: 21–23.

Asante, K. and Associates (1993) *Marketing Function Under Small Scale Mining Project*, Accra: Gesellschaft für Technische Zusammenarbeit.

Ashley, R. (1987) 'The Geopolitics of Geopolitical Space: Towards a Critical Social Theory of International Politics', *Alternatives* 12: 403–434.

Askin, S. and C. Collins (1993) 'External Collusion with Kleptocracy: Can Zaire Recapture Its Stolen Wealth?', *Review of African Political Economy* 57: 72–85.

Assiri, A.M., R.A. Parsons and N. Perdikis (1990) 'A Comparative Analysis of Debt Rescheduling in Latin America and Sub-Saharan Africa', *Scandinavian Journal of Development Alternatives* IX, 2 and 3: 117–127.

Atkinson, A. (1991) *Principles of Political Ecology*, London: Belhaven.

Auty, R. (1995) *Patterns of Development: Resources, Policy and Economic Growth*, London: Edward Arnold.

Azam, J.-P. (1994) 'The Uncertain Distribution Impact of Structural Adjustment in Sub-Saharan Africa', in R. Van Der Hoeven and F. Van Der Kraaij (eds) *Structural Adjustment and Beyond in Sub-Saharan Africa*, London and Portsmouth, NH: James Currey and Heinemann, 100–114.

Babb, F. (1996) 'After the Revolution: Neoliberal Policy and Gender in Nicaragua', *Latin American Perspectives* 23, 1: 27–48.

Balassa, B. (1981) *Structural Adjustment Policies in Developing Economies*, Washington, DC: World Bank.

Banuri, B.T. (1991) 'Introduction', in T. Banuri (ed.) *Economic Liberalization: No Panacea – The Experiences of Latin America and Asia*, Oxford: Clarendon Press.

Barya, J.J. (1993) 'The New Political Conditionalities of Aid: An Independent View from Africa', *IDS Bulletin* 24, 1: 16–23.

Bates, R.H. (1994) 'The Impulse to Reform in Africa', in J.A. Winder (ed.) *Economic Change and Political Liberalisation in Sub-Saharan Africa*, Baltimore and London: The Johns Hopkins University Press, 13–28.

Bayart, J.-F. (1991) 'Finishing with the Idea of the Third World: The Concept of Political Trajectory', in J. Manor (ed.) *Rethinking Third World Politics*, London: Longman.

—— (1993) *The State in Africa: The Politics of the Belly*, London: Longman.

Baylies, C. (1995) 'Political Conditionality and Democratisation', *Review of African Political Economy* 65: 321–337.

Beckman, B. (1992) 'Empowerment or Repression? The World Bank and the Politics of African Adjustment', in P. Gibbon *et al.* (eds) *Authoritarianism, Democracy and Adjustment: The Politics of Economic Reform in Africa*, Uppsala: Nordiska Afrikainstitutet.

—— (1993) 'The Liberation of Civil Society: Neo-liberal Ideology and Political Theory', *Review of African Political Economy* 58: 20–33.

Bello, W. (1989) 'Confronting the Brave New World Order: Toward a Southern Agenda for the 1990s', *Alternatives* 14: 135–167.

Bello, W. and S. Rosenfeld (1990) *Dragons in Distress: Asia's Miracle Economies in Crisis*, London: Penguin Books.

Bendiner, B. (1987) *International Labour Affairs: The World Trade Unions and the Multinational Companies*, Oxford: Clarendon Press.

Benton, T. and M. Redclift (1994) 'Introduction', in M. Redclift and T. Benton (eds) *Social Theory and the Global Environment*, London: Routledge.

Bernal, R. (1984) 'The IMF and Class Struggle in Jamaica, 1977–1980', *Latin American Perspectives* 11, 3: 53–82.

Bezanson, K. (1998) 'Tiger Cubs at the Crossroads – Some Policy Issues Facing Vietnam', *IDS Bulletin* 29, 3: 43–52.

Bienen, H. and J. Waterbury (1989) 'The Political Economy of Privatization in Developing Countries', *World Development* 17, 5: 617–632.

Bing, A. (1984) 'Popular Participation Versus People's Power: Notes on Politics and Power Struggles in Ghana', *Review of African Political Economy* 31: 91–104.

Bird, G. (1994) 'Changing Partners: Perspectives and Policies of the Bretton Woods Institutions', *Third World Quarterly* 15, 3: 483–503.

Blaikie, P. (1985) *The Political Economy of Soil Erosion in Developing Countries*, London: Longman.

—— (1995) 'Political Ecology for Developing Countries', *Geography* 80, 3: 203–214.

Blaikie, P. and H. Brookfield (1987) 'Defining and Debating the Problem', in P. Blaikie and H. Brookfield (eds) *Land Degradation and Society*, London: Methuen.

Blokland, K. (1996) 'Neoliberalism and the Central American Peasantry', in A. Fernandez Jilberto and A. Mommen (eds) *Liberalization in the Developing World: Institutional and Economic Changes in Latin America, Africa and Asia*, London: Routledge.

Boahen, A. (1989) *The Ghanaian Sphinx: Reflections on the Contemporary History of Ghana, 1972–1987*, Accra: Ghana Academy of Arts and Sciences.

Booth, D. (1985) 'Marxism and Development Sociology: Interpreting the Impasse', *World Development* 13, 7: 761–787.

Brand, V., R. Mupedziswa and P. Gumbo (1993) 'Women Informal Sector Workers in Zimbabwe', in P. Gibbon (ed.) *Social Change and Economic Reform in Africa*, Uppsala: Nordiska, 270–306.

Brecher, J. and T. Costello (1994) *Global Village or Global Pillage: Economic Reconstruction from the Bottom Up*, Boston, MA: South End Press.

Brett, E. (1988) 'Adjustment and the State: The Problem of Administrative Reform', *IDS Bulletin* 19, 4: 4–11.

Brown, E. (1996a) 'Nicaragua: Sandinistas, Social Transformation and the Continuing Search for a Popular Economic Programme', *Geoforum* 273: 275–295.

—— (1996b) 'Articulating Opposition in Latin America: The Consolidation of Neoliberalism and the Search for Radical Alternatives', *Political Geography* 15, 2:169–192.

Bruno, M. (1988) 'Opening Up: Liberalization and Stabilization', in R. Dornbusch and F. Helmers (eds) *The Open Economy: Tools for Policymakers in Developing Countries*, New York: Oxford University Press.

Bryant, R. (1992) 'Political Ecology: An Emerging Research Agenda in Third World Studies', *Political Geography* 11, 1: 12–36.

Bryant, R. and S. Bailey (1997) *Third World Political Ecology*, London: Routledge.

Brydon, L. and K. Legge (1995) 'Gender and Adjustment: Pictures from Ghana', in G.T. Emeagwali (ed.) *Women Pay the Price: Structural Adjustment in Africa and the Caribbean*, Trenton, New Jersey: Africa World Press, 63–86.

Buchanan, K. (1973) 'The White North and the Population Explosion', *Antipode* 5, 3: 7–15.

Burbach, R. (1997) 'Anniversary Essay: Socialism is Dead; Long Live Socialism', *NACLA Report on the Americas* 31, 3: 15–20.

Burbach, R., O. Nunez and B. Kagarlitsky (1997) *Globalization and Its Discontents: The Rise of Postmodern Socialisms*, London: Pluto Press.

Bush, R. (1997) 'Africa's Environmental Crisis: Challenging the Orthodoxies', *Review of African Political Economy* 74: 503–513.

Bush, R. and M. Szeftel (1994) 'Commentary: States, Markets and Africa's Crisis', *Review of African Political Economy* 60: 147–156.

—— (1995) 'Taking Leave of the Twentieth Century', *Review of African Political Economy* 65: 291–300.

Cahn, J. (1996) 'Challenging the New Imperial Authority: The World Bank and the Democratization of Development', *Harvard Human Rights Law Journal* 6: 159–194.

Callaghy, T. (1990) 'Lost Between State and Market: The Politics of Economic Adjustment in Ghana, Zambia and Nigeria', in J. Nelson, *Economic Crisis and Policy Choice: The Politics of Adjustment in the Third World*, Princeton: Princeton University Press.

Callaghy, T.M. and J. Ravenhill (eds) (1993) *Hemmed In: Responses to Africa's Economic Decline*, New York: Columbia University Press.

Cameron, J. (1992) 'Adjusting Structural Adjustment: Getting Beyond the UNICEF Compromise', in P. Mosley (ed.) *Development Finance and Policy Reform: Essays in the Theory and Practice of Conditionality in Less Developed Countries*, London: St Martin's Press.

Cannon, T. (1994) 'China's Reforms: Putting Their Money Where Their Mao Was', *Third World Quarterly* 15, 3: 522–527.

Cardoso, F. and E. Faletto (1979) *Dependency and Development in Latin America*, Berkeley: California University Press.

Carmody, P. (1998) 'Constructing Alternatives to Structural Adjustment in Africa', *Review of African Political Economy* 75: 25–46.

Cassen, R. (1994) 'Structural Adjustment in Sub-Saharan Africa', in W. Van Der Geest (ed.) *Negotiating Structural Adjustment in Africa*, London and Portmouth, NH: John Currey and Heinemann.

Cassen, R. and Associates (1994) *Does Aid Work?*, 2nd edition, Oxford: Oxford University Press.

Castañeda, J. (1993) *Utopia Unarmed: The Latin American Left after the Cold War*, New York: Vintage Books.

Cavarozzi, M. (1993) 'The Left in South America: Politics as the Only Option', in M. Vellinga (ed.) *Social Democracy in Latin America: Prospects for Change*, Boulder, Co.: Westview, 146–162.

Chachage, C. (1993) 'New Forms of Accumulation in Tanzania: The Case of Gold Mining', in C. Chachage, M. Ericsson and P. Gibbon, *Mining and Structural Adjustment*, Uppsala: Scandinavian Institute for African Studies.

Chalker, L. (1993) 'The Proper Role of Government', in D. Rimmer, *Action in Africa: The Experience of People Actively Involved in Government, Business and Aid*, London: James Currey.

Channel Four (1998) *The Bank, the President and the Pearl of Africa*, London: Channel Four/IBT.

Cheng, T. and S. Haggard (1987) *Newly Industrializing Asia in Transition: Policy Reform and American Response*, Policy Papers in International Affairs, 31, Berkeley: University of California.

Chin, C. and J. Mittelman (1997) 'Conceptualising Resistance to Globalisation', *New Political Economy* 2, 1: 25–37.

Chisari, O.O., J.M. Fanelli and R. Frenkel (1996) 'Argentina: Growth Resumption, Sustainability and Environment', *World Development* 24, 2: 227–240.

Chossudovsky, M. (1997) *The Globalization of Poverty: Impacts of IMF and World Bank Reforms*, London: Zed Books.

Chowdhury, A. and I. Islam (1993) *The Newly Industrialising Economies of East Asia*, London: Routledge.

Colburn, F. (1994) *The Vogue of Revolution in Poor Countries*, Princeton, NJ: Princeton University Press.

Cole, K. (1996) 'Cuba: The Options', mimeograph.

Conyers, D. and P. Hills (1984) *An Introduction to Development Planning in the Third World*, Chichester: John Wiley.

Copans, J. (1983) 'The Sahelian Drought: Social Sciences and the Political Economy of Underdevelopment', in K. Hewitt (ed.) *The Interpretation of Calamity*, London: Allen & Unwin.

Corbridge, S. (1986) *Capitalist World Development*, London: Macmillan.

—— (ed.) (1995) *Development Studies: A Reader*, London: Edward Arnold.

Cornia, G. (1998) 'Convergence on Governance, Dissent on Economic Policies', *IDS Bulletin* 29, 2: 32–38.

Cornia, G., R. Jolly and F. Stewart (1987) *Adjustment with a Human Face, Volumes 1 and 2, Protecting the Vulnerable and Promoting Growth*, Oxford: Clarendon Press.

Cowen, M. and R. Shenton (1996) *Doctrines of Development*, London: Routledge.

Cox, R. (1999) 'Civil Society at the Turn of the Millennium: Prospects for an Alternative World Order', *Review of International Studies* 25: 3–28.

Cromwell, E. (1996) 'Case Study for Tanzania', in D. Reed (ed.) *Structural Adjustment, the Environment, and Sustainable Development*, London: Earthscan.

Crush, J. (ed.) (1995) *Power of Development*, London: Routledge.

Culpeper, R. (1996) 'Multilateral Development Banks: Towards a New Division of Labour', in J. Griesgraber and B. Gunter (eds) *The World Bank: Lending on a Global Scale (Rethinking Bretton Woods Vol. 3)*, London: Pluto Press.

Daddieh, C. (1995) 'Education Adjustment and Under Severe Recessionary Pressures: the Case of Ghana', in K. Mengisteab and B.I. Logan (eds) *Beyond Economic Liberation in Africa: Structural Adjustment and the Alternatives*, London: Zed Books Limited, 23–55.

Danaher, K. (1994) *50 Years is Enough: The Case against the World Bank and the International Monetary Fund*, Boston, MA: South End Press.

Dasgupta, B. (1998) *Structural Adjustment, Global Trade and the New Political Economy of Development*, London: Zed Books.

Davidson, B. (1992) *The Black Man's Burden: Africa and the Curse of the Nation-State*, London: James Currey.

Davis, J. and C. Bishop (1998/9) 'The MAI: Multilateralism from Above', *Race and Class* 40, 2/3: 159–171.

De Soto, H. (1989) *The Other Path: The Invisible Revolution in the Third World*, New York: Harper & Row.

Dia, M. (1993) *A Governance Approach to Civil Service Reform in Sub-Saharan Africa – World Bank Technical Paper No. 225*, Washington: World Bank.

Dicken, P., J. Peck and A. Tickell (1997) 'Unpacking the Global', in R. Lee and J. Wills (eds) *Geographies of Economies*, London: Arnold, 158–166.

Dietz, J.L. (1987) 'Debt, International Corporation and Economic Change in Latin America and the Caribbean', *Latin American Perspectives* 55, 14, 4, Fall: 508–515.

Dolinsky, G.T. (1990) 'Debt and Structural Adjustment in Central America', *Latin American Perspectives* 67, 17, 4, Fall: 76–90.

Dornbusch, R. (1988) 'Balance of Payments Issues', in R. Dornbusch and F. Helmers (eds) *The Open Economy: Tools for Policymakers in Developing Countries*, New York: Oxford University Press.

Downs, A. (1957) *An Economic Theory of Democracy*, New York: Harper & Row.

Drake, P. (1991) 'Comment', in R. Dornbusch and S. Edwards (eds) *The Macroeconomics of Populism in Latin America*, Chicago: University of Chicago Press, 35–40.

Dzakpazu, V. (1992) *A Study of the Methods Used in Mining and Processing in a Sample of Small-scale Mining Operations*, Accra: Gesellschaft für Technische Zusammenarbeit.

Edwards, S. (1995) *Crisis and Reform in Latin America: From Despair to Hope*, Oxford: Oxford University Press.

Ehrlich, P. (1968) *The Population Bomb*, New York: Ballantine Books.

Ellner, S. (1993) 'Introduction: The Changing State of the Latin American Left in the Recent Past', in B. Carr and S. Ellner (eds) *The Latin American Left: From the Fall of Allende to Perestroika*, Boulder, Co. Westview, 1–22.

Elson, D. (1989) 'The Impact of Structural Adjustment on Women: Concepts and Issues', in B. Onimode (ed.) *The IMF, the World Bank and the African Debt: The Social and Political Impact*, London: Zed Books.

—— (1994) 'People, Development and International Financial Institutions: An Interpretation of the Bretton Woods System', *Review of African Political Economy* 62: 511–524.

—— (1995) 'Household Responses to Stabilisation and Structural Adjustment: Male Bias at the Micro Level', in D. Elson (ed.) *Male Bias in the Development Process*, Manchester and New York: Manchester University Press.

Emeagwali, G. (ed.) (1995) *Women Pay the Price: Structural Adjustment in Africa and the Caribbean*, Trenton, New Jersey: Africa World Press.

Engberg-Pedersen, P.E., P. Gibbon, P. Raikes and L. Udsholt (1996) *Limits of Adjustment in Africa*, Portsmouth, NH and Oxford: James Currey and Heinemann.

Engels, F. (1959) *Anti-Dühring*, Moscow: Foreign Languages Publishing House.

—— (1972) *Dialectics of Nature*, Moscow: Progress Publishers.

Englebert, P. and R. Hoffman (1996) *Burundi Learning the Lessons*, in I. Husain and R. Faruquee, *Adjustment in Africa: Lessons from Country Case Studies*, Washington, DC: World Bank.

Escobar, A. (1992) 'Imagining a Post-Development Era? Critical Thought, Development and Social Movements', *Social Text* 31/32: 20–56.

—— (1995) *Encountering Development: The Making and Unmaking of the Third World*, Princeton: Princeton University Press.

—— (1996) 'Constructing Nature: The Elements of a Poststructural Political Ecology', in R. Peet and M. Watts (eds) *Liberation Ecologies: Environment, Development and Social Movements*, London: Routledge.

Escobar, A. and S. Alvarez (eds) (1992) *The Making of Social Movements in Latin America: Identity, Strategy and Democracy*, Boulder, Co.: Westview.

Espinal, R. (1995) 'Economic Restructuring, Social Protest and Democratisation in the Dominican Republic', *Latin American Perspectives* 22, 3: 53–79.

Evans, P., D. Rueschemeyer and T. Skocpol (eds) (1985) *Bringing the State Back In*, Cambridge: Cambridge University Press.

Fagen, R., C. Deere and J. Corragio (1986) 'Introduction', in R. Fagen, C. Deere and J. Corragio, *Transition and Development: Problems of Third World Socialism*, New York: Monthly Review.

Falk, R. (1997) 'Resisting "Globalisation-from-above" Through "Globalisation-from-below"', *New Political Economy* 2, 1: 17–24.

Falloux, F. and L. Talbot (1993) *Crisis and Opportunity: Environment and Development in Africa*, London: Earthscan.

Falola, T. and J. Ihonvbere (1985) *The Rise and Fall of Nigeria's Second Republic, 1979–84*, London: Zed Books.

Fanelli, J., R. Frenkel and L. Taylor (1994) 'Is the Market Friendly Approach Friendly to Development? A Critical Assessment', in G. Bird and A. Helwege (eds) *Latin America's Economic Future*, San Diego, CA: Academic Press.

Fashoyin, T. (1994) 'Nigeria: Consequences for Employment', in A. Adepoju (ed.) *The Impact of Structural Adjustment on the Population of Africa*, Portsmouth, NH and London: James Currey and Heinemann.

Felix, D. (1998) 'Is the Drive Towards Free Market Globalization Stalling?', *Latin American Research Review* 33, 3: 191–216.

Ferguson, T. (1988) *The Third World and Decision Making in the International Monetary Fund: The Quest for Full and Effective Participation*, London and New York: Pinter Publishers.

Fernando, J. (1997) 'Nongovernmental Organizations, Micro-credit and the Empowerment of Women', *Annals of the American Academy of Political and Social Science* 554: 150–177.

Fine, B. (1999) 'The Developmental State is Dead – Long Live Social Capital?', *Development and Change* 30: 1–19.

Fitzgerald, E. and M. Wuyts (eds) (1988) *The Market Within Planning: Socialist Economic Management in the Third World*, London: Frank Cass.

Forrest, J. (1988) 'The Quest for State "Hardness" in Africa', *Comparative Politics* 20, 4: 423–442.

Frank, A.G. (1978) *Dependent Accumulation and Underdevelopment*, London and Basingstoke: The Macmillan Press Ltd.

Frankman, M. (1992) 'Global Income Redistribution: An Alternative Perspective on the Latin American Debt Crisis', in A. Ritter, M. Cameron and D. Pollock (eds) *Latin America to the Year 2000: Reactivating Growth, Improving Equity, Sustaining Democracy*, New York: Praeger.

Friedman, M. and R. Friedman (1980) *Free To Choose: A Personal Statement*, London: Secker & Warburg.

Fukuyama, F. (1992) *The End of History and the Last Man*, New York: Free Press.

Galenson, W. (1989) *The International Labour Organisation: An American View*, Madison: University of Wisconsin Press.

Gibbon, P. (1992) 'The World Bank and African Poverty, 1973–91', *The Journal of Modern African Studies* 30, 2: 193–220.

—— (1996) 'Structural Adjustment and Structural Change in Sub-Saharan Africa: Some Provisional Conclusions', *Development and Change* 27: 751–784.

Gibbon, P. and Y. Bangura (1993) *Social Change and Economic Reform in Africa*, Uppsala: Nordiska Afrikainstitutet.

Gibbon, P., Y. Bangura and A. Ofstad (eds) (1994) *Authoritarianism, Democracy and Adjustment: The Politics of Economic Reform in Africa*, Uppsala: Nordiska Afrikainstitutet.

Giddens, A. (1993) *The Consequences of Modernity*, Cambridge: Polity Press.

Gillies, D. (1996) 'Human Rights, Democracy and Good Governance: Stretching the World Bank's Policy Frontiers', in J. Greisgraber and B. Gunter (eds) (1996) *The World Bank: Lending on a Global Scale (Rethinking Bretton Woods Vol. 3)*, London: Pluto Press.

Gills, B. (1997) 'Editorial: "Globalisation" and the "Politics of Resistance"', *New Political Economy* 2, 1: 11–15.

Gills, B. and J. Rocamora (1992) 'Low Intensity Democracy', *Third World Quarterly* 13, 3: 501–523.

Glyn, A. and R. Sutcliffe (1992) 'Global But Leaderless? The New Capitalist Order', *Socialist Register*, London: Merlin Press.

Gordon, D.F. (1993) 'Debt, Conditionality, and Reform: The International Relations of Economic Policy Restructuring in Sub-Saharan Africa', in T.M. Callaghy and J. Ravenhill (eds) *Hemmed In: Responses to Africa's Economic Decline*, New York: Columbia University Press.

Graham, C. (1992) 'The Politics of Protecting the Poor During Adjustment: Bolivia's Emergency Social Fund', *World Development* 20, 9: 1,233–1,251.

Green, D. (1995) *The Silent Revolution: The Rise of Market Economics in Latin America*, London: Cassell.

——(1996) 'Latin America: Neoliberal Failure and the Search for Alternatives', *Third World Quarterly* 17, 1: 109–122.

Greenaway, D. and O. Morrissey (1993) 'Structural Adjustment and Liberalisation in Developing Countries: What Lessons Have We Learned?', *Kyklos* 46: 241–261.

Griesgraber, J.M. and B.G. Gunter (eds) (1996a) *Development: New Paradigms and Principles for the Twenty-First Century (Rethinking Bretton Woods Vol. 2)*, London: Pluto Press.

—— (eds) (1996b) *The World Bank: Lending on a Global Scale (Rethinking Bretton Woods Vol. 3)*, London: Pluto Press.

Griffin, K. and J. Gurley (1985) 'Radical Analyses of Imperialism, the Third World and the Transition to Socialism: A Survey Article', *Journal of Economic Literature* 23: 1,089–1,143.

Grosh, B. (1994) 'Through the Structural Adjustment Minefield: Politics in an Era of Economic Liberation', in J.A. Widner (ed.) *Economic Change and Political Liberation in Sub-Saharan Africa*, Baltimore and London: The Johns Hopkins University Press, 29–46.

Haggard, S. (1990) *Pathways from the Periphery: The Politics of Growth in Newly Industrializing Countries*, Ithaca: Cornell University Press.

Hague, R., M. Harrop and S. Breslin (1992) *Comparative Government and Politics: An Introduction*, London: Macmillan.

Hajari, N. (1997) 'Dark Cloud of Death', *Time* 6 October: 40–43.

Hammond, J. (1995) 'Book Review: A Farewell to Arms? Review of J. Castañeda, *Utopia Unarmed*', *Latin American Perspectives* 22, 4: 115–120.

Harris, R. (1992) *Marxism, Socialism and Democracy in Latin America*, Boulder, Co.: Westview.

Harrison, P. (1993) *The Third Revolution: Population, Environment and a Sustainable World*, London: Penguin.

Harvey, D. (1974) 'Population, Resources and the Ideology of Science', *Economic Geography* 50.

Hatem, M.V. (1994) 'Privatisation and the Demise of State Feminism in Egypt', in P. Sparr (ed.) *Mortgaging Women's Lives*, London: Zed Books, 40–60.

Hawthorn, G. and P. Seabright (1996) 'Governance, Democracy and Development: A Contractualist View', in A. Leftwich (ed.) *Democracy and Development*, Cambridge: Polity Press.

Hayter, T. (1971) *Aid as Imperialism*, Harmondsworth: Penguin.

Hayter, T. and C. Watson (1985) *Aid-Rhetoric and Reality*, London: Pluto Press Ltd.

Hearn, J. (1998) 'The NGO-isation of Kenyan Society: USAID and the Restructuring of Healthcare', *Review of African Political Economy* 25: 89–100.

Held, D. (1989) 'The Decline of the Nation State', in S. Hall and M. Jacques (eds) *New Times: The Changing Face of Politics in the 1990s*, London: Lawrence & Wishart.

—— (1992) 'Democracy: From City-States to a Cosmopolitan Order?', *Political Studies* 40.

Hellman, J. (1997) 'Social Movements: Revolution, Reform and Reaction', *NACLA Report on the Americas* 30, 6: 13–18.

Henstridge, M. (1994) 'Stabilisation Policy and Structural Adjustment in Uganda', in W. Van der Geest, *Negotiating Structural Adjustment in Africa*, London and Portsmouth, NH: James Currey and Heinemann, 197–222.

Herbold Green, R. (1998) 'A Cloth Untrue: The Evolution of Structural Adjustment in Sub-Saharan Africa', *Journal of International Affairs* 52, 1: 207–232.

Herbold Green, R. and M. Faber (1994) 'Editorial: The Structural Adjustment of Structural Adjustment: Sub-Saharan Africa 1980–1993', *IDS Bulletin* 25, 3: 1–8.

Hibou, B. (1999) 'The "Social Capital" of the State as an Agent of Deception', in J. Bayart, S. Ellis and B. Hibou, *The Criminalization of the State in Africa*, The International Africa Institute in association with James Currey (Oxford) and Indiana University Press (Bloomington).

Hildyard, N. and A. Wilks (1998) 'An Effective State? But Effective for Whom?', *IDS Bulletin* 29, 2: 49–55.

Hirst, P. (1997) 'The Global Economy – Myths and Realities', *International Affairs* 73, 3: 409–425.

Hirst, P. and G. Thompson (1996) *Globalization in Question: The International Economy and the Possibilities of Governance*, Cambridge: Polity Press.

Hodgin, G. (1998) 'Socialism against Markets: A Critique of Recent Proposals', *Economy and Society* 27, 4: 407–433.

Hollaway, J. (1996) 'Environmental Problems in Zimbabwe from Gold Panning', *CRS Perspectives* 52: 25–28.

Hoogvelt, A. (1997) *Globalisation and the Postcolonial World: The New Political Economy of Development*, Basingstoke: Macmillan.

Hopkins, A.G. (1973) *An Economic History of West Africa*, London: Longman.

Hulme, D. and M. Edwards (1997) 'NGOs, States and Donors: An Overview', in D. Hulme and M. Edwards (eds) *NGOs, States and Donors: Too Close for Comfort*, Basingstoke: Macmillan.

Husain, I. and R. Faruquee (1995) *Adjustment in Africa: Lessons from Country Case Studies*, Washington, DC: World Bank.

Hutchful, E. (1989) 'From "Revolution" to Monetarism: The Economics and Politics of Structural Adjustment in Ghana', in B. Campbell and J. Loxley (eds) *Structural Adjustment in Africa*, London: Macmillan.

Hutton, W. (1986) *The Revolution that Never Was: An Assessment of Keynesian Economics*, Harlow: Longman.

Hyden, G. (1983) *No Shortcuts to Progress: African Development Management in Perspective*, London: Heinemann.

——(1997) 'Civil Society, Social Capital, and Development: Dissection of a Complex Discourse', *Studies in Comparative International Development* 32, 1: 3–30.

Ihonvbere, J. and O. Vaughan (1995) 'Democracy and Civil Society: The Nigerian Transition Programme, 1985–1993', in J. Wiseman (ed.) *Democracy and Political Change in Sub-Saharan Africa*, London: Routledge.

Imber, M. (1997) 'Geo-governance Without Democracy? Reforming the UN System', in A. McGrew (ed.) *The Transformation of Democracy?*, Cambridge: Polity/Open University.

IMF (1997) *International Financial Statistics Yearbook 1997*, Washington: International Monetary Fund.

Jameson, K. and C. Wilber (1981) 'Socialism and Development: Editor's Introduction', *World Development* 9, 9/10.

Jessop, B. (1990) *State Theory: Putting Capitalist States in Their Place*, Cambridge: Polity Press.

Johnson, O.E.G. (1994) 'Managing Adjustment Costs, Political Authority and the Implementation of Adjustment Programmes with Special Reference to African Countries', *World Development* 22, 3: 399–411.

Jolly, R. (1990) 'Poverty and Adjustment in the 1990s', *African Environment* 7, 1–4: 239–253.

Jonah, K. (1989) 'The Social Impact of Ghana's Adjustment Programme, 1983–86', in B. Onimode (ed.) *The IMF, the World Bank and the African Debt, Volume 2: The Social and Political Impact*, London and New Jersey: Zed Books.

Jordan, A. (1995) 'Designing New International Organizations: A Note on the Structure and Operation of the Global Environment Facility', *Public Administration* 73: 303–312.

Judson, F. (1993) 'The Making of Central American National Agendas under Adjustment and Restructuring', *Labour, Capital and Society* 26, 2: 148–180.

Jumah, A. (1993) *Environmental Audit to Study the Effects of Small-scale Mining Activities on Vegetation*, Accra: Gesellschaft für Technische Zusammenarbeit.

Kandeh, J. (1992) 'Sierra Leone: Contradictory Class Functionality of the "Soft" State', *Review of African Political Economy* 55: 30–43.

Kaplan, R. (1994) 'The Coming Anarchy: How Scarcity, Crime, Overpopulation, and Disease are Rapidly Destroying the Social Fabric of the Planet', *Atlantic Monthly* 273, 2: 44–77.

Kay, C. (1989) *Latin American Theories of Development and Underdevelopment*, London: Routledge.

Kay, G. (1975) *Development and Underdevelopment: A Marxist Analysis*, London: Macmillan.

Kiely, R. (1998) 'Neo Liberalism Revised? A Critical Account of World Bank Concepts of Good Governance and Market Friendly Intervention', *Capital and Class* 64: 63–88.

Killick, T. (1995) 'Structural Adjustment and Poverty Alleviation: An Interpretative Survey', *Development and Change* 26: 305–331.

—— (1996) 'Principals, Agents and the Limitations of BWI Conditionality', *The World Economy* 19, 2: 211–229.

—— (1999) 'Making Adjustment Work for the Poor', *ODI Poverty Briefing* 5 May: 1–4.

Killick, T. and C. Stevens (1991) 'Eastern Europe: Lessons on Economic Adjustment from the Third World', *International Affairs* 67, 4: 679–696.

Killick, T., P. Gunaltilaka and A. Marr (1998) *Aid and the Political Economy of Policy Change*, London: Routledge.

Kirkpatrick, C. and Z. Onis (1991) 'Turkey', in P. Mosley, J. Harrigan and J. Toye (eds) *Aid and Power: The World Bank and Policy-Based Lending, Volume 2, Case Studies*, London: Routledge.

Kirkpatrick, C. and J. Weiss (1995) 'Trade Policy Reforms and Performance in Africa in the 1980s', *The Journal of Modern African Studies* 33, 2: 285–298.

Kirton, C.D. (1989) 'Grenada and the IMF: The PRG's Extended Fund Facility Programme, 1983', *Latin American Perspectives* 62, 16, 3, Summer: 121–144.

Kitching, G. (1982) *Development and Underdevelopment in Historical Perspective: Populism, Nationalism and Industrialisation*, London: Methuen.

Kothari, R. (1984) 'Party and State in our Times: The Rise of Non-party Political Formations', *Alternatives* 9: 542.

Kreuger, A. (1974) 'The Political Economy of the Rent-Seeking Society', *The American Economic Review* 64, 3: 291–303.

Kwon, J. (1994) 'The East Asia Challenge to Neo-Classical Orthodoxy', *World Development* 22: 635–644.

Lal, D. (1983) *The Poverty of 'Development Economics'*, London: Institute of Economic Affairs.

Lancaster, C. (1993) 'Governance and Development: The Views from Washington', *IDS Bulletin* 24, 1: 9–15.

Lander, E. (1996) 'The Impact of Neoliberal Adjustment in Venezuela, 1989–1993', *Latin American Perspectives* 90, 23, 3, Summer: 50–73.

Laurie, N. (1997) 'From Work to Welfare: The Response of the Peruvian State to the Feminisation of Emergency Work', *Political Geography* 16, 8: 691–714.

Leechor, C. (1996) *Ghana: Frontrunner in Adjustment*, in I. Husain and R. Faruquee *Adjustment in Africa: Lessons from Country Case Studies*, Washington, DC: World Bank.

Leftwich, A. (1993) 'Governance, Democracy and Development in the Third World', *Third World Quarterly* 14, 3: 605–624.

—— (1994) 'Governance, the State and the Politics of Development', *Development and Change* 25: 363–386.

—— (1995) 'Bringing Politics Back In: Towards a Model of the Developmental State', *The Journal of Development Studies* 31, 3: 400–427.

—— (1996) 'On the Primacy of Politics in Development', in A. Leftwich (ed.) *Democracy and Development*, Cambridge: Polity Press.

Lehman, H. (1993) *Indebted Development: Strategic Bargaining and Economic Adjustment in the Third World*, Basingstoke: Macmillan.

Lehmann, D. (1990) *Democracy and Development in Latin America: Economics, Politics and Religion in the Post-War Period*, Cambridge: Polity Press.

Lele, S. (1991) 'Sustainable Development: A Critical Review', *World Development* 19, 6: 607–621.

Leys, C. (1994) 'Confronting the African Tragedy', *New Left Review* 204: 33–47.

—— (1996) *The Rise and Fall of Development Theory*, London: James Currey.

Lipschutz, R. (1992) 'Reconstructing World Politics: The Emergence of Global Civil Society', *Millennium* 21, 3: 389–420.

Little, I., R. Cooper, W. Max Corden and S. Rajapatirana (1993) *Boom, Crisis, and Adjustment: The Macroeconomic Experience of Developing Countries*, Oxford University Press/World Bank.

Logan, B.I. (1995) 'Can Sub-Saharan Africa Successfully Privatise Its Health Care?', in K. Mengisteab and I. Logan (eds) *Beyond Economic Liberation in Africa: Structural Adjustment and the Alternatives*, London: Zed Books, 56–74.

Logan, I. and K. Mengisteab (1993) 'IMF–World Bank Adjustment and Structural Transformation in Sub-Saharan Africa', *Economic Geography* 69, 1: 1–24.

London Environment and Economics Centre (1993) 'Case Study for Mexico', in D. Reed (ed.) *Structural Adjustment and the Environment*, London: Earthscan.

Loxley, J. (1990) 'Structural Adjustment in Africa: Reflections on Ghana and Zambia', *Review of African Political Economy* 47, Spring: 8–27.

—— (1992) *Ghana: The Long Road to Recovery, 1983–1990*, Ottawa: The North–South Institute.

Loxley, J. and D. Seddon (1994) 'Stranglehold on Africa', *Review of African Political Economy* 62: 485–493.

Lugalla, J.L.P. (1993) 'Structural Adjustment Policies and Education in Tanzania', in P. Gibbon (ed.) *Social Change and Economic Reform in Africa*, Uppsala: Nordiska Afrikainstitutet, 184–214.

Malima, K, (1994) 'Structural Adjustment: The African Experience', in R. Van Der Hoeven and F. Van Der Kraaij (eds) *Structural Adjustment and Beyond in Sub-Saharan Africa*, London and Portsmouth, NH: James Currey and Heinemann, 9–16.

Manor, J. (1995) 'Politics and the Neo-liberals', in C. Colclough and J. Manor (eds) *States or Markets? Neo-liberalism and the Development Policy Debate*, Oxford: Clarendon Press.

Manuh, T. (1994) 'Ghana: Women in the Public and Informal Sectors under the Economic Recovery Programme', in P. Sparr (ed.) *Mortgaging Women's Lives: Feminist Critique of Structural Adjustment*, London: Zed Books, 61–77.

Marcussen, H. (1996) 'NGOs, the State and Civil Society', *Review of African Political Economy* 69: 405–423.

Markandya, A. (1996) 'Case Study for Jamaica', in D. Reed (ed.) *Structural Adjustment, the Environment, and Sustainable Development*, London: Earthscan.

Marr, A. (1999) *The Poor and Their Money: What Have We Learned?*, ODI Poverty Briefing, no. 4, March, London: ODI.

Marshall, D. (1996) 'National Development and the Globalisation Discourse: Confronting "Imperative" and "Convergence" Notions', *Third World Quarterly* 17, 5: 875–901.

Marshall, J. (1998) 'The Political Viability of Free Market Experimentation in Cuba: Evidence from Los Mercados Agropecuarios', *World Development* 26, 2: 277–288.

Marshall, R. (1991) 'Power in the Name of Jesus', *Review of African Political Economy* 52: 21–37.

Martin, M. (1993) 'Neither Phoenix Nor Icarus: Negotiating Economic Reform in Ghana and Zambia, 1983–1992', in T.M. Callaghy and J. Ravenhill (eds) *Hemmed In: Responses to Africa's Economic Decline*, New York: Columbia University Press, 130–179.

Martinussen, J. (1997) *Society, State and Market: A Guide to Competing Theories of Development*, London: Zed Books.

—— (1998) 'The Limitations of the World Bank's Conception of the State and the Implications for Institutional Development Strategies', *IDS Bulletin* 29, 2: 67–74.

Mbaku, M. (1996) 'Bureaucratic Corruption in Africa: The Futility of Cleanups', *Cato Journal* 16, 1: 99–118.

MacDonald, L. (1995a) 'A Mixed Blessing: The NGO Boom in Latin America', *NACLA Report on the Americas* 28, 5: 30–35.

—— (1995b) 'NGOs and the Problematic Discourse of Participation: Cases from Costa Rica', in D. Moore and G. Schmitz (eds) *Debating Development Discourse: Institutional and Popular Perspectives*, Basingstoke: Macmillan.

—— (1997) *Supporting Civil Society: The Political Role of Non-governmental Organizations in Central America*, Basingstoke: Macmillan.

McGrew, A. (1997) 'Democracy Beyond Borders? Globalization and the Reconstruction of Democratic Theory and Practice', in A. McGrew (ed.) *The Transformation of Democracy?*, Cambridge: Polity Press.

McIlwaine, K. (1998) 'Civil Society and Development Geography', *Progress in Human Geography* 22, 3: 415–424.

Meadows, D., D. Meadows, J. Randers and W. Behrens (1972) *The Limits to Growth*, London: Pan.

Mearns, R. (1991) *Environmental Implications of Structural Adjustment: Reflections on Scientific Method*, IDS Discussion Paper No. 284, Brighton: IDS.

Meiksins Wood, E. (1995) *Democracy Against Capitalism*, Cambridge: Cambridge University Press.

Mesa-Lago, C. (1997) 'Social Welfare Reform in the Context of Economic-Political Liberalization: Latin American Cases', *World Development* 25, 4: 497–517.

Mihevc, J. (1995) *The Market Tells Them So*, London: Zed Books/Third World Network.

Miliband, R. (1996) 'The New World Order and the Left', *Social Justice* 23, 1–2: 14–20.

Minerals Commission (1991) *Guidelines (Environmental) for Draft Regulations on Mining in Ghana*, Accra.

Mohan, G. (1994) 'Manufacturing Consensus: (Geo)political Knowledge and Policy-based Lending', *Review of African Political Economy* 62: 525–538.

—— (1996) 'Decentralisation and Adjustment in Ghana: A Case of Diminished Sovereignty', *Political Geography* 15, 1: 75–94.

Moore, M. (1998) 'Toward a Useful Consensus?', *IDS Bulletin* 29, 2: 39–48.

Mosley, P. (1994) 'Decomposing the Effects of Structural Adjustment: The Case of Sub-Saharan Africa', in R. Van De Hoeven and F. Van Der Kraaij (eds) *Structural Adjustment and Beyond in Sub-Saharan Africa*, Portsmouth, NH and London: James Currey and Heinemann, 70–98.

Mosley, P., J. Harrigan and J. Toye (eds) (1991) *Aid and Power: The World Bank and Policy-Based Lending: Volume One: Analysis and Policy Proposals*, London: Routledge.

Mosley, P., J. Hudson and S. Horrell (1987) 'Aid, the Public Sector and the Markets in Less Developed Countries', *Economic Journal* 97, September: 616–641.

Mosley, P., T. Subasat and J. Weeks (1995) 'Assessing Adjustment in Africa', *World Development* 23, 9: 1,459–1,473.

Munck, R. (1994) 'Workers, Structural Adjustment, and Concertacion Social in Latin America', *Latin American Perspectives* 21, 3: 90–103.

Mustapha, A.R. (1992) 'Structural Adjustment and Multiple Modes of Social Livelihood in Nigeria', in P. Gibbon, Y. Bangura, and A. Ofstad (eds) *Authoritarianism, Democracy and Adjustment: The Politics of Economic Reform in Africa*, Uppsala: Nordiska Afrikainstitutet, 188–216.

Nabli, M. and J. Nugent (1989) 'The New Institutional Economics and Its Applicability to Development', *World Development* 17, 9: 1,333–1,347.

NACLA (North American Congress on Latin America) (1993) 'The Latin American Left: A Painful Rebirth', *NACLA Report on the Americas* 25, 5: 12.

—— (1995) 'Cuba: Adapting to a Post-Soviet World', *NACLA Report on the Americas* 29, 2: 6.

—— (1997) 'Voices on the Left: A Thirtieth Anniversary Celebration', *NACLA Report on the Americas* 31, 1: 5–6.

Nafziger, E.W. (1993) *The Debt Crisis in Africa*, Baltimore and London: Johns Hopkins University Press.

Najam, A. (1998) 'Searching for NGO Effectiveness', *Development Policy Review* 16, 3: 305–310.

Nazmi, N. (1997) 'Exchange-Rate-Based Stabilisation in Latin America', *World Development* 25, 4: 519–535.

Nelson, J. (1992) 'Good Governance: Democracy and Conditional Economic Aid', in P. Mosley (ed.) *Development Finance and Policy Reform: Essays in the Theory and Practice of Conditionality in Less Developed Countries*, London: St Martin's Press.

Norton, A. and T. Stephens (1995) *Participation in Poverty Assessments*, Social Policy and Resettlement Division, Washington, DC: The World Bank.

NSR (1991) 'Study on the Effect of Mining on the Environment' (draft), Victoria, Australia.

Nyamekye, E. (1992) 'Environmental Management in the Small-scale Mining Industry in Ghana', in P. Acquah (ed.) *Seminar on the Effect of Mining on Ghana's Environment with Particular Reference to Proposed Mining Environmental Guidelines*, Accra: Minerals Commission.

O Tuathail, G. (1996) *Critical Geopolitics: The Politics of Writing Global Space*, London: Routledge.

Obi, C. (1997) 'Globalisation and Local Resistance: The Case of the Ogoni versus Shell', *New Political Economy* 2, 1: 137–148.

Ofei-Aboagye, E. (1992) *A Study of the Participation of Women in Small-scale Mining*, Accra: Deutsche Gesellschaft für Technische Zusammenarbeit.

Ohmae, K. (1990) *The Borderless World*, London: Collins.

Olowu, D. and P. Smoke (1992) 'Determinants of Success in African Local Governments: An Overview', *Public Administration and Development* 12: 1–17.

Olson, M. (1965) *The Logic of Collective Action*, Cambridge, MA: Harvard University Press.

Olukoshi, A.O. (1989) 'Impacts of IMF–World Bank Programmes on Nigeria' in B. Onimode (ed.) *The IMF, the World Bank and the African Debt (Volume 1: The Economic Impact)*, London: Zed Books.

—— (1994a) 'Introduction', in A.O. Olukoshi, R.O. Olaniyan and F. Aribisala (eds) *Structural Adjustment in West Africa*, Lagos, Nigeria: Pumark, 1–10.

—— (1994b) 'Structural Adjustment in West Africa: An Overview', in A.O. Olukoshi, R.O. Olaniyan and F. Aribisala (eds) *Structural Adjustment in West Africa*, Lagos, Nigeria: Pumark, 28–38.

Oluyemi-Kusa, A. (1994) 'The Structural Adjustment Programme of the Nigerian State', in A.O. Olukoshi *et al.* (eds) *Structural Adjustment in West Africa*, Lagos: Pumark.

Onis, Z. (1995) 'The Limits of Neoliberalism: Toward a Reformulation of Development Theory', *Journal of Economic Issues* 39, 1: 97–119.

Organisation for Economic Cooperation and Development (1993) *Economic Instruments for Environmental Management in Developing Countries*, Paris: OECD.

Orme Jr, W.A. (1993) 'The Accelerating Pace of Privatization in Latin America', in P.H. Boeker (ed.) *Latin America's Turnaround: Privatisation, Foreign Investment and Growth*, San Francisco: Institute for Contemporary Studies.

Osaghae, E.E. (1995) *Structural Adjustment and Ethnicity*, Uppsala: Nordiska Afrikainstitutet, Research Report No. 98.

Overbeek, H. and K. Van der Pijl (1993) 'Restructuring Capital and Restructuring Hegemony: Neo-Liberalism and the Unmaking of the Post-war Order', in H. Overbeek (ed.) *Restructuring Hegemony in the Global Political Economy: The Rise of Transnational Neoliberalism in the 1980s*, London: Routledge.

Oxfam (1994) *Embracing the Future... Avoiding the Challenge of World Poverty* (mimeo), Oxford, September 1994.

Parfitt, T. (1990) 'Lies, Damned Lies and Statistics: The World Bank/ECA Controversy', *Review of African Political Economy* 47: 129–141.

Parfitt, T. and S. Riley (1989) *The African Debt Crisis*, London: Routledge.

Pastor, M. and A. Zimbalist (1995) 'Cuba's Economic Conundrum', *NACLA Report on the Americas* 29, 2: 7–12.

Pearce, J. (1997) 'Between Co-option and Irrelevance? Latin American NGOs in the 1990s', in D. Hulme and M. Edwards (eds) *NGOs, States and Donors: Too Close for Comfort*, Basingstoke: Macmillan.

Peel, Q. and S. Thoenes (1997) 'Smoke at the End of the World', *Financial Times* 18 October: 7.

Peet, R. and M. Watts (eds) (1996) *Liberation Ecologies: Environment, Development and Social Movements*, London: Routledge.

Petras, J. (1997a) 'Alternatives to Neoliberalism in Latin America', *Latin American Perspectives* 24,1: 80–91.

—— (1997b) 'Latin America: The Resurgence of the Left', *New Left Review* 233: 17–47.

Pieterse, J. (1997) 'Globalisation and Emancipation: From Local Empowerment to Global Reform', *New Political Economy* 2, 1: 79–92.

Pio, A. (1992) 'The Social Dimension of Economic Adjustment Programmes: Economic Feedbacks and Implications for Medium and Long-Term Growth', in P. Mosley (ed.) *Development Finance and Policy Reform: Essays in the Theory and Practice of Conditionality in Less Developed Countries*, London: St Martin's Press.

Polanyi, K. (1960) *The Great Transformation: The Political and Economic Origins of Our Time*, Boston: Beacon Press.

Portantiero, J. (1993) 'Foundations of a New Politics', *NACLA Report on the Americas* 25, 5: 17–20.

Post, K. and P. Wright (1989) *Socialism and Underdevelopment*, London: Routledge.

Potter, R., T. Binns, J. Elliott and D. Smith (1999) *Geographies of Development*, Harlow: Longman.

Preobrazhensky, A.I. (1965) *The New Economics*, Oxford: Clarendon Press.

Preston, P. (1996) *Development Theory: An Introduction*, Oxford: Blackwell.

Pugh, C. (1997) 'Viewpoint: The World Bank's Millennial Theory of the State: Further Attempts to Reconcile the Political and the Economic', *Third World Planning Review* 19, 3: iii–xiv.

Radice, H. (1996) 'The Question of Globalization: A Review of Hirst and Thompson', paper presented at the 26th Annual Conference of Socialist Economists, University of Northumbria at Newcastle, 12–14 July.

Rahman, A. (1999) 'Micro-credit Initiatives for Equitable and Sustainable Development: Who Pays?', *World Development* 27, 1: 67–82.

Ramirez, M. (1993) 'Stabilization and Adjustment in Latin America: A Neostructuralist Perspective', *Journal of Economic Issues* 4: 1,015–1,040.

Randall, V. and R. Theobald (1985) *Political Change and Underdevelopment: A Critical Introduction*, London: Macmillan.

Rapley, J. (1994) 'New Directions in the Political Economy of Development', *Review of African Political Economy* 62: 495–510.

Ravenhill, J. (1993) 'A Second Decade of Adjustment: Greater Complexity, Greater Uncertainty', in T.M. Callaghy and J. Ravenhill (eds) *Hemmed In: Responses to Africa's Economic Decline*, New York: Columbia University Press, 18–53.

Ray, D. (1986) *Ghana: Politics, Economics and Society*, London: Pinter.

Redclift, M. (1984) *Development and the Environmental Crisis: Red or Green Alternatives*, London: Routledge.

—— (1987) *Sustainable Development: Exploring the Contradictions*, London: Routledge.

—— (1995) 'The Environment and Structural Adjustment: Lessons for Policy Interventions in the 1990s', *Journal of Environmental Management* 44: 55–68.

Redclift, M. and T. Benton (eds) (1994) *Social Theory and the Global Environment*, London: Routledge.

Reed, D. (ed.) (1993) *Structural Adjustment and the Environment*, London: Earthscan.

—— (ed.) (1996) *Structural Adjustment, the Environment, and Sustainable Development*, London: Earthscan.

Republic of Ghana (1987) *Programme of Actions to Mitigate the Social Costs of Adjustment*, Accra: Republic of Ghana.

—— (1993) *Quarterly Digest of Statistics*, Statistical Service, Accra.

Rich, B. (1994) *Mortgaging the Earth: The World Bank, Environmental Impoverishment and the Crisis of Development*, London: Earthscan.

Richards, P. (1996) *Fighting for the Rain Forest: War, Youth and Resources in Sierra Leone*, Oxford and Portsmouth, NH: James Currey and Heinemann.

Richardson, D. (1983) *The Role and Significance of the International Labour Organisation*, Trent Business School Open Lectures on Industrial Relations.

Rist, G. (1997) *The History of Development: From Western Origins to Global Faith*, London: Zed Books.

Roberts, K. (1995) 'Neoliberalism and the Transformation of Populism in Latin America: The Peruvian Case', *World Politics* 48: 82–116.

Robinson, W. (1999) 'Latin America and Global Capitalism', *Race and Class* 40, 2–3: 111–131.

Rocha, J. (1999) 'San Francisco Libre: Giving It One More Try', *Envio* 210–211: 36–46.

Rondinelli, D., J.S. McCullough and R.W. Johnson (1989) 'Analyzing Decentralization Policies in Developing Countries: A Political-Economy Framework', *Development and Change* 20: 57–87.

Ros, J., J. Draisma, N. Lustig and A.T. Kate (1996) 'Prospects for Growth and the Environment in Mexico in the 1990s', *World Development* 24, 2: 307–324.

Rosen, F. (1997) 'Back on the Agenda: 10 Years After the Debt Crisis', *NACLA Report on the Americas* 31, 3: 21–24.

Roth, M. (1991) 'Structural Adjustment in Perspective: Challenges for Africa in the 1990s', in I. Deng, M. Kostner and C. Young (eds) *Democratisation and Structural*

Adjustment in Africa in the 1990s, African Studies Programme: University of Wisconsin – Madison.

Rothchild, D. (1993) 'Rawlings and the Engineering of Legitimacy in Ghana', paper presented to the conference on elections in Ghana at the University of London, August.

Rothchild, D. and N. Chanzan (eds) (1988) *The Precarious Balance: State and Society in Africa*, Boulder, Co.: Westview.

Rothchild, D. and M. Foley (1983) 'The Implications of Scarcity for Governance in Africa', *International Political Science Review* 4, 3: 311–326.

Rothchild, D. and E. Gyimah-Boadi (1988) 'Ghana's Economic Decline and Development Strategies', in J. Ravenhill (ed.) *Africa in Economic Crisis*, London: Macmillan.

Rouis, M. (1996) 'Senegal: Stabilization, Partial Adjustment, and Stagnation', in I. Husain and R. Faruquee (eds) *Adjustment in Africa: Lessons from Country Case Studies*, Washington, DC: World Bank.

Ruccio, D. (1991) 'When Failure Becomes Success: Class and the Debate over Stabilization and Adjustment', *World Development* 19, 10: 1,315–1,334.

Safa, H.I. (1995) 'Economic Restructuring and Gender Subordination', *Latin American Perspectives* 85, 22, 2, Spring: 32–50.

Sahn, D.E., P.A. Dorosh and S.D. Younger (1994) *Economic Reform in Africa: A Foundation for Poverty Alleviation*, Cornell Food and Nutrition Policy Programme, September, Working Paper, 72.

Schumpeter, J. (1934) *The Theory of Economic Development*, London: Cambridge University Press.

Schuurman, F. (ed.) (1993) *Beyond the Impasse: New Directions in Development Theory*, London: Zed Books.

Scott, G.E. (1992) 'Structural Adjustment in Sub-Saharan Africa: A Review of IMF Stabilisation Programs', in K.E. Bauzon (ed.) *Development and Democratisation in the Third World: Myths, Hopes, and Realities*, London: Taylor & Francis.

Sebastian, I. and A. Alicbusan (1989) *Sustainable Development: Issues in Adjustment Lending Policies*, Environment Department Divisional Working Papers 1989–6, Washington, DC: World Bank.

Seckler, D. and R. Cobb (1991) *African Development: Lessons from Asia*, Arlington, Virginia: Winrock International Institute for Agricultural Development, 13–22.

Seshamani, V. (1994) 'Structural Adjustment and Poverty Alleviation: Some Issues on the Use of Social Safety Nets and Targeted Public Expenditures', in R. Van Der Hoeven and F. Van Der Kraaij (eds) *Structural Adjustment and Beyond in Sub-Saharan Africa*, London and Portsmouth, NH: James Currey and Heinemann, 114–125.

Shaw, L. (nd) *Social Clauses*, Catholic Institute for International Relations Briefing, London: CIIR.

Shefner, J. (1995) 'Moving in the Wrong Direction in Social Movement Theory', *Theory and Society* 24: 595–612.

Simon, D. (1995) 'The Demise of Socialist State Forms in Africa: An Overview', *Journal of International Development* 7, 5: 707–739.

Sinclair, M. (ed.) (1995) *The New Politics of Survival: Grassroots Movements in Central America*, New York: Monthly Review.

Singer, H. (1991) 'Terms of Trade: New Wine and New Bottles', *Development Policy Review* 9, 4: 339–352.

Singh, K., N. Dawkins-Scully and D. Wysham (1997) 'Micro Credit: Band-aid or Wound?', available on the Internet at http://web.m-web.com/kav001.html.

Sklar, R. (1996) 'Towards a Theory of Developmental Democracy', in A. Leftwich (ed.) *Democracy and Development*, Cambridge: Polity Press.

Slater, D. (1989) 'Territorial Power and the Peripheral State: The Issue of Decentralization', *Development and Change* 20: 501–531.

—— (1993) 'The Geopolitical Imagination and the Enframing of Development Theory', *Transactions of the Institute of British Geographers* (new series) 18: 419–437.

Solodovnikov, V. and V. Bogoslovsky (1975) *Non-Capitalist Development: An Historical Outline*, Moscow: Progress.

Sparr, P. (ed.) (1994) *Mortgaging Women's Lives: Feminist Critiques of Structural Adjustment*, London: Zed Books.

Stein, H. (1994) 'The World Bank and the Application of Asian Industrial Policy to Africa: Theoretical Considerations', *Journal of International Development* 6, 3.

Stern, E. (1993) 'World Bank Financing of Structural Adjustment', in J. Williamson (ed.) *IMF Conditionality*, Boston: MIT Press.

Stewart, F. (1991) 'The Many Faces of Adjustment', *World Development* 19, 12: 1,847–1,864.

—— (1999) 'Biases in Global Markets: Can the Forces of Inequity and Marginalization be Modified', in M. ul Haq, R. Jolly, P. Streeton and K. Haq (eds) (1999) *The UN and the Bretton Woods Institutions: New Challenges for the 21st Century*, Basingstoke: Macmillan.

Stewart, S. (1997) 'Happy Ever After in the Market Place: Non-government Organisations and Uncivil Society', *Review of African Political Economy* 71: 11–34.

Svejnar, J. and K. Terrell (1988) *Industrial Labor: Enterprise Ownership and Government Policies*, Pittsburgh: University of Pittsburgh Press.

Swamy, G. (1996) 'Kenya: Patchy, Intermittent Commitment', in I. Husain and R. Faruqee (eds) *Adjustment in Africa: Lessons from Country Case Studies*, Washington, DC: The World Bank.

Synge, R. (1989) 'Nigeria to 1993: Will Liberalisation Work?', *The Economist Intelligence Unit*, Special Report No. 1,134, EIU Economic Prospects Series, 15.

Tarp, F. (1993) *Stabilization and Structural Adjustment: Macroeconomic Frameworks for Analysing the Crisis in Sub-Saharan Africa*, London: Routledge.

Taylor, L. (1997) 'Editorial: The Revival of the Liberal Creed – the IMF and the World Bank in a Globalized Economy', *World Development* 25, 2: 145–152.

Teivainen, T. (1994) *The International Monetary Fund: A Modern Priest – The Politics of Economism and the Containment of Changes in the Global Political Community*, paper presented at the XVIth World Congress of the International Political Science Association, 21–25 August, Berlin.

Teriba, O. (1996) 'The Challenge of Africa's Socio-Economic Transformation', in J. Griesgraber and B. Gunter (eds) *The World Bank: Lending on a Global Scale (Rethinking Bretton Woods Vol. 3)*, London: Pluto Press.

Tevera, D. (1995) 'The Medicine that Might Kill the Patient: Structural Adjustment and Urban Poverty in Zimbabwe', in D. Simon, W. Van Spengen, C. Dixon, A. Narman (eds), *Structurally Adjusted Africa: Poverty, Debt and Basic Needs*, London and Boulder, Co.: Pluto Press, 79–90.

Thoenes, S. (1997) 'Indonesia Given Promises of Help to Fight Fires', *Financial Times* 26 September: 4.

Toye, J. (1993) *Dilemmas of Development*, Oxford: Blackwell.

—— (1994) 'Structural Adjustment: Context, Assumptions, Origin and Diversity', in R. Van Der Hoeven and F. Van Der Kraaij (eds) *Structural Adjustment and Beyond in Sub-Saharan Africa*, London and Portsmouth, NH: James Currey and Heinemann, 18–35.

—— (1995) 'Is There a New Political Economy of Development?', in C. Colclough and J. Manor (eds) *States or Markets? Neo-liberalism and the Development Policy Debate*, Oxford: Clarendon Press, 321–338.

Toye, J. and C. Jackson (1996) 'Public Expenditure Policy and Poverty Reduction: Has the World Bank got it Right?', *IDS Bulletin* 27, 1: 56–66.

Umoden, G. (1992) *The Babangida Years*, Lagos: Gabumo Publishing Company.

UNRISD (1995) *States of Disarray: The Social Effects of Globalization*, Geneva: UNRISD.

Utting, P. (1992) *Economic Reform and Third World Socialism: A Political Economy of Food Policy in Post-Revolutionary Societies*, New York: St Martin's Press.

Van de Walle, N. (1993) 'Political Liberalisation and Economic Policy Reform in Africa', in *Economic Reform in Africa's New Era of Political Liberalisation*, Proceedings of a Workshop For SPA Donors, USAID, 14–15 April, Washington, DC.

Van der Geest, W. and A. Kottering (1994) 'Structural Adjustment in Tanzania: Objectives and Achievements', in W. van der Geest (ed.) *Negotiating Structural Adjustment in Africa*, London: James Currey.

Van Dormael, A. (1978) *Bretton Woods: Birth of a Monetary System*, London and Basingstoke: The Macmillan Press.

Veltmeyer, H. (1997) 'New Social Movements in Latin America: The Dynamics of Class and Identity', *The Journal of Peasant Studies* 25, 1: 139–169.

Veltmeyer, H., J. Petras and S. Vieux (1997) *Neoliberalism and Class Conflict in Latin America: A Comparative Perspective on the Political Economy of Structural Adjustment*, Basingstoke: Macmillan.

Vidal, J. (1997) 'When the Earth Caught Fire', *The Guardian* 8 November.

Vilas, C. (1990) 'Las Economias Periféricas Frente a la Transforación Revolucionaria: El Caso de Nicaragua', *Cuadernos de Pensamiento Propio: Series Ensayos* 18: 125–145.

—— (1992) 'What Future for Socialism?', *NACLA Report on the Americas* 25, 5: 13–16.

Walters, A. (1994) *Do We Need the IMF and the World Bank?*, London: Institute of Economic Affairs.

Walton, J. (1995) 'Urban Protest and the Global Political Economy: The IMF Riots', in M.P. Smith and J.R. Feagin (eds) *The Capitalist City: Global Restructuring and Community Politics*, Oxford: Basil Blackwell.

Wapenhans, W. (1994) 'The Political Economy of Structural Adjustment: An External Perspective', in R. Van Der Hoeven and F. Van Der Kraaij (eds) *Structural Adjustment and Beyond in Sub-Saharan Africa*, London and Portsmouth, NH: James Currey and Heinemann, 36–52.

Warren, B. (1980) *Imperialism: Pioneer of Capitalism*, London: Verso.

Waterman, P. (1996) 'Review Article: Beyond Globalism and Developmentalism: Other Voices in World Politics', *Development and Change* 27: 165–180.

Waters, M. (1995) *Globalization*, London: Routledge.

Westebbe, R. (1994) 'Structural Adjustment, Rent Seeking, and Liberalisation in Benin', in J.A. Widner (ed.) *Economic Change and Political Liberalisation in Sub-Saharan Africa*, Baltimore and London: Johns Hopkins University Press, 80–100.

Weyland, K. (1996) 'Neopopulism and Neoliberalism in Latin America: Unexpected Affinities', *Studies in Comparative International Development* 31, 3: 3–31.

White, G. (1983) 'Revolutionary Socialist Development in the Third World: An Overview', in G. White, R. Murray and C. White (eds) *Revolutionary Socialist Development in the Third World*, London: Wheatsheaf.

Wiarda, H. (1987) *Latin America at the Crossroads: Debt, Development and the Future*, Boulder, Co.: Westview.

Williams, D. and T. Young (1994) 'Governance, the World Bank and Liberal Theory', *Political Studies* 42: 84–100.

Williams, G. (1995) 'Modernizing Malthus: The World Bank, Population Control and the African Environment', in J. Crush (ed.) *Power of Development*, London: Routledge.

Winpenny, J. (1996) 'Case Study for Cameroon', in D. Reed (ed.) *Structural Adjustment, the Environment, and Sustainable Development*, London: Earthscan.

Woodhouse, P. (1997) 'Governance and Local Environmental Management in Africa', *Review of African Political Economy* 74: 537–547.

Woodward, D. (1992) *Debt, Adjustment and Poverty in Developing Countries: Volume One: National and International Dimensions of Debt and Adjustment in Developing Countries*, London: Pinter/Save the Children.

—— (1993) *Structural Adjustment Policies: What Are They? Are They Working?*, CIIR Briefing Papers: CIIR.

World Bank (1983) *World Development Report*, Washington, DC: World Bank.

—— (1988) *World Bank Development Report 1988: Opportunities and Risks in Managing the World Economy*, Oxford: Oxford University Press.

——(1989) *Sub-Saharan Africa: From Crisis to Sustainable Growth*, Washington, DC: World Bank.

—— (1990) *Strengthening Local Government in Sub-Saharan Africa*, Washington, DC: World Bank.

—— (1991) *World Development Report*, Washington, DC: World Bank.

—— (1992a) *Poverty Reduction Operational Directive*, Washington, DC: World Bank.

—— (1992b) *Strategy for African Mining*, Washington, DC: World Bank.

—— (1992c) *World Development Report*, Washington, DC: World Bank.

—— (1993a) *The East Asian Miracle*, Oxford: Oxford University Press.

—— (1993b) *Governance: The World Bank's Experience*, Operation's Policy Department, Washington, DC: World Bank.

—— (1995) *Small and Microenterprize Finance: Guiding Principles for Selecting and Supporting Intermediaries*, Washington, DC: World Bank.

—— (1997) *World Bank Development Report 1997: The State in a Changing World*, Oxford: Oxford University Press.

Wunsch, J. (1991) 'Institutional Analysis and Decentralization: Developing an Analytical Framework for Effective Third World Administrative Reform', *Public Administration and Development* 11: 431–451.

Zack-Williams, A.B. (1990) 'Sierra Leone: Crisis and Despair', *Review of African Political Economy* 49, Winter: 22–33.

—— (1991) 'The Politics of Crisis and Ethnicity in Sierra Leone', paper presented at the Centre for African Studies, University of Liverpool, February.

—— (1992) 'Sierra Leone: The Deepening Crisis and Survival Strategies', in J.E Nyang'oro and T.M. Shaw, *Beyond Structural Adjustment in Africa: The Political Economy of Sustainable and Democratic Development*, New York: Praeger, 149–168.

—— (1993) 'Crisis, Structural Adjustment and Creative Survival in Sierra Leone', *Africa Development* 18, 1: 53–66.

—— (1994) 'A Failed Industrial Revolution: Merchant Capital and Textile Consumption in West Africa', *Africa Technology Forum* 7, 2.

—— (1995a) *Tributors, Supporters and Merchant Capital: Mining and Underdevelopment in Sierra Leone*, Aldershot: Avebury.

—— (1995b) 'Crisis and Structural Adjustment in Sierra Leone: Implication for Women', in G.T. Emeagwali (ed.) *Women Pay the Price: Structural Adjustment in Africa and the Caribbean*, Trenton, New Jersey: Africa World Press, 53–62.

—— (1997) 'Labour, Structural Adjustment, and Democracy in Sierra Leone and Ghana', in R.A. Siddiqui (ed.) *Subsaharan Africa in the 1990s: Challenges to Democracy and Development*, Westport USA: Praeger, 57–69.

——(1999) 'The Political Economy of Civil War in Sierra Leone', *Third World Quarterly* 20, 1: 143–162.

Zuvekas, C. (1997) 'Latin America's Struggle for Equitable Economic Adjustment', *Latin American Research Review* 32, 2: 152–169.

Index

Printed and bound by CPI Group (UK) Ltd, Croydon, CR0 4YY

01/11/2024

01782621-0007